SMALL APPLIANCES

Other Publications:

THE TIME-LIFE GARDENER'S GUIDE

MYSTERIES OF THE UNKNOWN

TIME FRAME

FIX IT YOURSELF

FITNESS, HEALTH & NUTRITION

SUCCESSFUL PARENTING

HEALTHY HOME COOKING

UNDERSTANDING COMPUTERS

LIBRARY OF NATIONS

THE ENCHANTED WORLD

THE KODAK LIBRARY OF CREATIVE PHOTOGRAPHY

GREAT MEALS IN MINUTES

THE CIVIL WAR

PLANET EARTH

COLLECTOR'S LIBRARY OF THE CIVIL WAR

THE EPIC OF FLIGHT

THE GOOD COOK

WORLD WAR II

HOME REPAIR AND IMPROVEMENT

THE OLD WEST

SMALL APPLIANCES

TIME-LIFE BOOKS
ALEXANDRIA, VIRGINIA

Fix It Yourself was produced by
ST. REMY PRESS

MANAGING EDITOR	Kenneth Winchester
MANAGING ART DIRECTOR	Pierre Léveillé

Staff for *Small Appliances*

Series Editor	Kathleen M. Kiely
Editor	Elizabeth W. Lewis
Series Art Director	Diane Denoncourt
Art Director	Philippe Arnoldi
Research Editor	Kent J. Farrell
Designers	Maryse Doray, Solange Pelland
Editorial Assistant	Heather Mills
Contributing Writers	Margaret Caldbick, Chris Cockrill, Catherine Gerols, Carol Halls, Olivier Humber, Emer Killean, Michael Kleiza, Anita Malhotra, Charles Mappin, Daniel McBain, Rachel Wareham
Electronic Designer	Daniel Bazinet
Contributing Illustrators	Gérard Mariscalchi, Jacques Proulx
Technical Illustrators	Nicolas Moumouris, Robert Paquet
Cover	Robert Monté
Index	Christine M. Jacobs
Administrator	Denise Rainville
Coordinator	Michelle Turbide
Systems Manager	Shirley Grynspan
Systems Analyst	Simon Lapierre
Studio Director	Maryo Proulx

Time-Life Books Inc. is a wholly owned subsidiary of
TIME INCORPORATED

FOUNDER	Henry R. Luce 1898-1967
Editor-in-Chief	Jason McManus
Chairman and Chief Executive Officer	J. Richard Munro
President and Chief Operating Officer	N. J. Nicholas Jr.
Editorial Director	Ray Cave
Executive Vice President, Books	Kelso F. Sutton
Vice President, Books	George Artandi

TIME-LIFE BOOKS INC.

EDITOR	George Constable
Executive Editor	Ellen Phillips
Director of Design	Louis Klein
Director of Editorial Resources	Phyllis K. Wise
Editorial Board	Russell B. Adams Jr., Dale M. Brown, Roberta Conlan, Thomas H. Flaherty, Lee Hassig, Donia Ann Steele, Rosalind Stubenberg, Henry Woodhead
Director of Photography and Research	John Conrad Weiser
Asst. Director of Editorial Resources	Elise Ritter Gibson
PRESIDENT	Christopher T. Linen
Chief Operating Officer	John M. Fahey Jr.
Senior Vice Presidents	Robert M. DeSena, James L. Mercer
Vice Presidents	Stephen L. Bair, Ralph J. Cuomo, Neal Goff, Stephen L. Goldstein, Juanita T. James, Hallett Johnson III, Carol Kaplan, Susan J. Maruyama, Robert H. Smith, Joseph J. Ward
Director of Production Services	Robert J. Passantino

Editorial Operations

Copy Chief	Diane Ullius
Production	Celia Beattie
Library	Louise D. Forstall
Correspondents	Elizabeth Kraemer-Singh (Bonn); Maria Vincenza Aloisi (Paris); Ann Natanson (Rome).

THE CONSULTANTS

Consulting Editor **David L. Harrison** served as an editor for several Time-Life Books do-it-yourself series, including *Home Repair and Improvement*, *The Encyclopedia of Gardening* and *The Art of Sewing*.

Evan Powell is Director of Chestnut Mountain Research Inc. in Taylors-Greenville, South Carolina, a firm that specializes in the development and evaluation of home appliances. He is a contributing editor to several do-it-yourself magazines, and has written a number of books on appliance repair.

Ira Gladstone owns and operates the Mr. Fix-It Service Center in Montreal, Quebec. A second-generation small appliance repairman, he has worked in his family-owned repair business for more than 15 years.

Michael R. MacDonald is a technical director for television production and a freelance photographer. He is a regular consultant for *Fix-It-Yourself*.

Steve Toth writes a monthly home-appliance repair column for *Popular Mechanics*. He has worked as an appliance service technician and has taught appliance repair at a New Jersey technical school.

Steven J. Forbis was a writer and editor for Time-Life Books' *Home Repair and Improvement* series. He is an editor of *Prodigy Service*, an interactive electronic information and entertainment services network for home computer users.

Library of Congress Cataloging-in-Publication Data
Small appliances.
         p.   cm. – (Fix it yourself)
    Includes index.
    ISBN 0-8094-6256-7
    ISBN 0-8094-6257-5 (lib. bdg.)
1. Household appliances, Electric—Maintenance and repair—Amateurs' manuals.   I. Time-Life Books.   II. Series.
TK9901.S58   1988
643' .6—dc19   88-19922
                                                        CIP

For information about any Time-Life book, please write:
Reader Information
Time-Life Customer Service
P.O. Box C-32068
Richmond, Virginia
23261-2068

CONTENTS

HOW TO USE THIS BOOK

Small Appliances is divided into three sections. The Emergency Guide on pages 8-11 provides information that can be indispensable, even lifesaving, in the event of a household emergency. Take the time to study this section *before* you need the important advice it contains.

The Repairs section—the heart of the book—is a comprehensive approach to troubleshooting and repairing small, electrical household appliances. Pictured below are four sample pages from the chapter on mixers, with captions describing the various features of the book and how they work. If a governor-

type stand mixer doesn't run at all, for example, the Troubleshooting Guide will offer a number of possible causes. If the problem is a dirty or faulty governor switch, you will be directed to page 74 for detailed, step-by-step directions for cleaning, testing and replacing the governor switch.

Each job has been rated by degree of difficulty and by the average time it will take for a do-it-yourselfer to complete. Keep in mind that this rating is only a suggestion. Before deciding whether you should attempt a repair, first read all the instructions carefully. Then be guided by your own confidence and the

---

Introductory text
Describes working principles of the appliance, most common breakdowns and basic safety precautions.

Troubleshooting Guide
To use this chart, locate the symptom that most closely resembles your appliance's problem, review the possible causes in column 2, then follow the recommended procedures in column 3. Simple fixes may be explained on the chart; in most cases you will be directed to an illustrated, step-by-step repair sequence.

Exploded and cutaway diagrams
Locate and describe the various components of the appliance.

MIXERS

Using a variety of beaters and speed settings, an electric mixer can do everything from kneading bread dough to whipping egg whites. Most mixers have a universal motor that turns a worm gear, driving a pair of pinion gears in opposite directions from each other. When beaters are inserted into the spindles of the pinion gears, they create the opposing rotation that mixes food so efficiently.

A typical hand mixer is illustrated at right. Some hand mixers are cordless rechargeables; to determine the source of a problem with one of these, also review the section on rechargeables in Tools & Techniques *(page 122)*.

Most stand mixers look similar to the one pictured on page 72. This type uses a sensitive governor switch to control motor speed. When the control knob is rotated to the desired speed, it closes the switch contacts to start the motor. The rapid rotation

of the motor armature activates a thrust rod, which repeatedly opens and closes the switch contacts to maintain precise speed control. Some newer stand mixers have an electronic circuit board instead of a governor switch. The electronic type is illustrated on page 75.

To avoid mixer problems caused by misuse, first review your owner's manual. When the mixer does not work properly, consult the categories that apply to your mixer in the Troubleshooting Guide below.

If your mixer hums but doesn't work, or if it has a burning smell, turn it off at once to avoid damaging the motor or gears. If you suspect a faulty motor, consult Tools & Techniques *(page 123)*. Keep in mind that many mixer motors are not easily serviced, and replacing a motor can be more expensive than buying a new mixer.

TROUBLESHOOTING GUIDE

SYMPTOM	POSSIBLE CAUSE	PROCEDURE
ALL MIXERS		
Mixer doesn't run at all	No power to outlet or outlet faulty	Reset breaker or replace fuse (p. 112) □○; have outlet serviced
	Power cord faulty	Test and replace power cord (p. 116) ▨●▲
	Internal wiring faulty	Inspect and repair wire connections (p. 118) ▨○
Motor hums; beaters don't turn	Motor bearing seized	Service motor (p. 123) ▨●
Mixer overheats	Air intake vents clogged	Clean air intake vents with a cotton swab or vacuum cleaner
	Motor bearings worn or dry	Service motor (p. 123) ▨●▲
Mixer vibrates noisily	Gears faulty	Service gears (p. 77) ▨○
	Motor faulty	Service motor (p. 123) ▨●▲
Mixer gives electrical shock	Wire connections loose or broken	Repair wire connections (p. 118) ▨○
Beaters hit one another, jam, grind, fall out, or won't turn	Gears faulty	Service gears (p. 77) ▨○
	Beater shafts worn or bent	Inspect and replace beaters
HAND MIXERS		
Mixer doesn't run at all	Speed control or power burst switch faulty	Test and replace main switch assembly (p. 71) ▨●
	Motor faulty	Service motor (p. 123) ▨●▲
GOVERNOR-TYPE STAND MIXERS		
Mixer doesn't run at all	Governor switch faulty	Service governor switch assembly (p. 74) ▨○
	Motor faulty	Service motor (p. 123) ▨●▲
Mixer runs only on high speed, or doesn't turn off	Governor switch faulty	Service governor switch assembly (p. 74) ▨○
	Condenser faulty	Service condenser (p. 74) ▨○▲
Mixer runs sluggishly or erratically	Governor switch or resistor faulty	Service governor switch assembly (p. 74) ▨○
	Motor faulty	Service motor (p. 123) ▨●▲
ELECTRONIC STAND MIXERS		
Mixer doesn't run at all	On/off switch faulty	Service on/off switch (p. 76) ▨○
	Fuse or motor faulty	Test fuse and motor circuit (p. 76) ▨○▲; service fuse (p. 77) ▨○ or service motor (p. 123) ▨○▲ as required
	Circuit board faulty	Replace circuit board (p. 77) ▨○
Mixer runs only on high speed	Circuit board faulty	Replace circuit board (p. 77) ▨○

DEGREE OF DIFFICULTY: □ Easy ▨ Moderate ■ Complex
ESTIMATED TIME: ○ Less than 1 hour ◐ 1 to 3 hours ● Over 3 hours ▲ Special tool required

70

HAND MIXERS

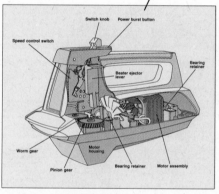

Switch knob   Power burst button

Speed control switch

Beater ejector lever

Bearing retainer

Worm gear

Motor housing

Pinion gear

Bearing retainer   Motor assembly

Accessing switches, motor and gears. Turn off and unplug the mixer. Remove the beaters and the detachable power cord, if it has one. Pry the switch knob off the lever. Turn over the mixer and remove the screws from the motor housing. Turn the mixer upright and work apart the housing, pulling the upper housing off over the beater ejector lever. You now have access to the motor and switch assembly. To access the gears, turn over the motor housing and pry off the small metal retaining rings and flat washers around the bottoms of the gear spindles, using a small screwdriver or long-nose pliers. Turn the motor housing upright. Unscrew the switch assembly from the top of the front bearing retainer and move the switch assembly to one side. You can now service the gears *(page 77)*.

SERVICING THE SWITCHES (Hand mixers)

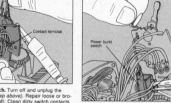

Power cord wire

Contact terminal

Power burst switch

Power cord wire

1 **Testing the speed control switch.** Turn off and unplug the mixer. Access the switches *(step above)*. Repair loose or broken wire connections *(page 118)*. Clean dirty switch contacts *(page 117)*. To test a speed control switch like the one shown here, use a continuity tester, or set a multitester to test continuity *(page 113)*. Identify the power cord wire connected to the switch and clip one tester probe to its terminal on the switch. Set the switch lever to the first contact terminal and touch the other probe to that terminal. The multitester should show continuity. Repeat the test at each switch lever setting, in turn *(above)*. If there is continuity at each setting, the switch is OK; go to step 3. If there is no continuity at a setting, replace the switch assembly *(step 3)*.

2 **Testing the power burst switch.** Use a continuity tester, or set a multitester to test continuity *(page 113)*. Identify the power cord wire connected to the power burst switch, located on the main switch assembly. Clip one tester probe to the wire's terminal on the switch. Then touch the second probe to one of the other two terminals as you depress the shaft of the power burst switch *(above)*. The multitester should show continuity. Repeat the test for the other terminal. If there is continuity, the switch is OK. If there is no continuity in a test, replace the switch assembly.

71

Degree of difficulty and time
Rate each repair by complexity and by how much time the job should take for a person with average do-it-yourself skills.

Special tool required
Some repairs require a specialized tool or a multitester *(page 110)*.

Cross-references
Direct you to important information elsewhere in the book, including disassembly and access steps.

tools and time available to you. For more complex or time-consuming repairs, such as servicing a motor, you may wish to take the mixer for professional service. You will still have saved time and money by diagnosing the problem yourself.

Most repairs in *Small Appliances* can be made with a set of screwdrivers, a continuity tester and a soldering iron. Any special tool required is indicated in the Troubleshooting Guide. Basic tools — and the proper way to use them — are presented in Tools & Techniques *(page 110)*. If you are a novice at home repair, read this section in preparation for a job.

Repairing a small appliance is easy and safe if you work logically and follow the tips and precautions. Before beginning a repair, turn off power to the appliance and unplug it. Set the unit on a clean work table and, as you disassemble it, write down the sequence of steps. Make a capacitor discharging tool *(page 114)* and discharge the capacitor if the appliance has one. Store fasteners and other small parts in labeled containers, and tag all wires before removing them from their terminals. Perform a cold check for leaking voltage *(page 114)* after reassembling the appliance but before plugging it back in.

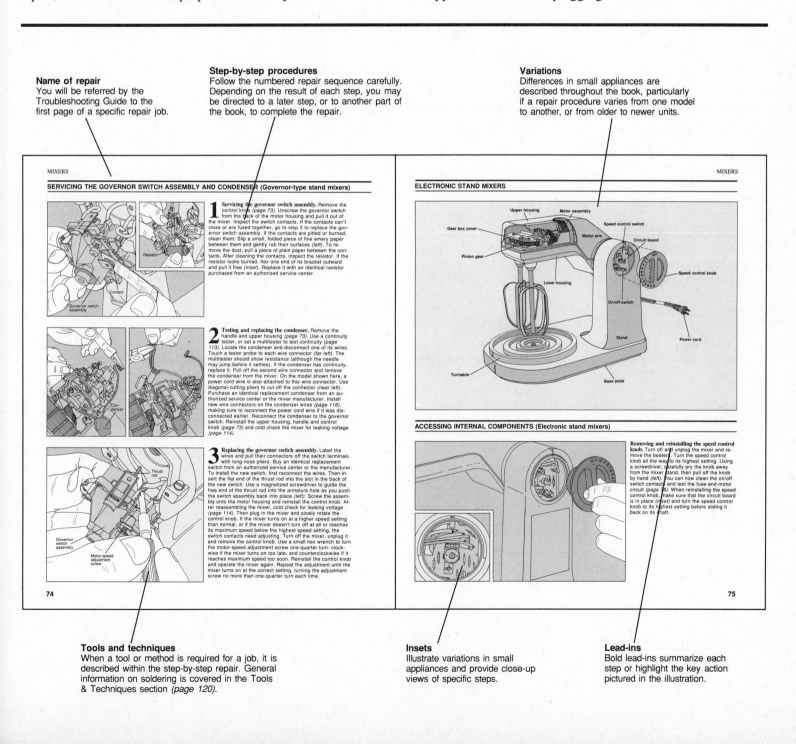

Name of repair
You will be referred by the Troubleshooting Guide to the first page of a specific repair job.

Step-by-step procedures
Follow the numbered repair sequence carefully. Depending on the result of each step, you may be directed to a later step, or to another part of the book, to complete the repair.

Variations
Differences in small appliances are described throughout the book, particularly if a repair procedure varies from one model to another, or from older to newer units.

Tools and techniques
When a tool or method is required for a job, it is described within the step-by-step repair. General information on soldering is covered in the Tools & Techniques section *(page 120)*.

Insets
Illustrate variations in small appliances and provide close-up views of specific steps.

Lead-ins
Bold lead-ins summarize each step or highlight the key action pictured in the illustration.

EMERGENCY GUIDE

Preventing small appliance problems. A faulty appliance usually creates a minor, temporary annoyance—the iron won't heat up, or the toaster won't latch down. But on rare occasions, a failure in its mechanical or electrical system can pose a real danger. With proper use and care, a small appliance will usually last, accident-free, for years longer than its warranty. A careful review of the appliance's owner's manual and its chapter in this book will help to prevent unnecessary breakdowns and—more importantly—safety hazards. Several manufacturers offer toll-free consumer "hot lines" for answering questions immediately and they will mail you an owner's manual free upon request. Their telephone numbers are available from the 800 operator.

The Troubleshooting Guide on page 9 puts emergency procedures for small appliance mishaps at your fingertips. It lists the quick-action steps to take, and refers you to the procedures on pages 10 and 11 for more detailed instructions. Read the emergency instructions thoroughly before you need to use them, so no time is lost in the event of a real emergency. Post the telephone numbers for the fire department, your utility companies and medical emergency services by the telephone; in most areas, dial 911 in case of a life-threatening emergency.

While small appliances make many of life's more mundane tasks simpler, they often have heating elements that can burn and blades and fans that can cut. Never leave a plugged-in appliance unattended or accessible to small children. Dangling power cords, and the fascinating shapes and sounds of appliances, can make these potentially hazardous machines seem like toys to children. Tips on using small appliances safely can be found on page 11.

Electrical shock and fire are life-threatening emergencies that can happen in even the most safety-conscious of homes. Deprive fire of its sneak attack by installing smoke alarms judiciously throughout the house. Have the correct fire extinguisher on hand to snuff out a blaze before it gets the upper hand *(page 10)*. If you must rescue someone stuck to an appliance by live current, do not touch the person; use a wooden implement of any kind to push the victim free *(page 10)*.

The repair of small appliances need not be any more dangerous than their daily use. In fact, proper repairs will prevent hazardous conditions caused by wear and neglect. The list of safety tips at right covers basic guidelines for safe service and use of any small appliance. See the chapters on individual small appliances for more specific advice, and review the information provided in Tools and Techniques *(page 110)* so that you will know how to proceed with repairs safely.

SAFETY TIPS

1. Familiarize yourself with the owner's manual for each appliance. If you have misplaced the manual, purchase one from an authorized service center or from the manufacturer.

2. Before attempting any repair in this book, read the entire procedure. Familiarize yourself with the specific safety information in each chapter.

3. Always turn off and unplug a small appliance and allow it to cool before cleaning or servicing it.

4. Never run a small appliance on an extension cord with an amperage rating less than the appliance's. Do not use an extension cord as permanent wiring for a small appliance.

5. Never allow small children to play with or operate small appliances. If children are nearby, do not leave working appliances unattended. Store small appliances out of the reach of children when not in use.

6. Do not submerge a small appliance in water to clean it.

7. Never use small appliances in or near the bath or shower. Do not leave small appliances where they can fall into water.

8. Never reach for a plugged-in small appliance that has fallen into water. Turn off the power at the main service panel, then unplug the appliance from the wall outlet before retrieving it.

9. Do not stick hands or utensils into any appliance with a blade mechanism while it is plugged in or turned on. Handle sharp appliance blades and cutters with care.

10. Do not poke a fork or other metal implement inside a toaster, hair dryer or any other small appliance with a heating element while it is plugged in or turned on.

11. Install a ground fault circuit interrupter (GFCI) on each circuit used by small bathroom and kitchen appliances to minimize the risk of electrical shock.

12. Label the breakers or fuses on your main service panel so you know which outlets they control and what their amperage is. Quickly shut off power to an outlet if an appliance begins sparking or burning.

13. When resetting a circuit breaker or replacing a fuse at the main service panel, work with one hand, holding the other hand behind you or using it to grasp a plastic flashlight.

14. Install smoke detectors and fire extinguishers in your home.

15. When repairing a small appliance, never bypass or alter any switch or component. Do not remove the ground prong of a three-prong power cord.

16. When replacing a faulty small appliance part, bring the part with you and buy an identical replacement from a manufacturer-authorized service center. Look for a UL (Underwriters Laboratories) or CSA (Canadian Standards Association) seal.

17. Always cold check for leaking voltage *(page 114)* before using an appliance for the first time after a repair.

TROUBLESHOOTING GUIDE

PROBLEM	PROCEDURE
Appliance gives electrical shock	Shut off power at main service panel, then unplug appliance from wall outlet
	Locate and repair cause of shock, or take appliance for service
Appliance is excessively hot	Shut off power at main service panel, then unplug appliance from wall outlet
	Clean the appliance as instructed in the owner's manual and this book, or take it for service
Power cord or plug is sparking or excessively hot	Shut off power at main service panel, then unplug appliance from wall outlet
	Test and replace power cord *(p. 116)*
Fire in appliance or outlet	Use fire extinguisher rated for electrical fires *(p. 9)*, and call fire department
	Shut off power at main service panel, then unplug appliance from wall outlet
	If flames or smoldering continue, leave house and wait for fire department
Fire inside toaster oven or microwave oven	Do not open door of appliance. Shut off power at main service panel, then unplug appliance from wall outlet
Smoke or burning odor coming from appliance	Shut off power at main service panel, then unplug appliance from wall outlet. Take appliance for service
Appliance falls into water while plugged in	Do not touch appliance or water. Shut off power at main service panel, then unplug power cord. Take appliance for service
Person immobilized by electrical shock from appliance	Unplug appliance if you can do so without touching victim. Otherwise, push victim away from appliance with a wooden implement *(p. 10)*
	Check whether victim is breathing and has a pulse. If not, call for medical help immediately. Administer artificial resuscitation or cardiopulmonary resuscitation (CPR), if qualified. Otherwise, place victim in recovery position *(p. 10)* until medical help arrives
Burn or scald from hot appliance	Soak injury in cold water. Do not apply any ointment or butter. If severe, cover with sterile gauze and seek medical help
Skin cut by sharp appliance parts	Wrap cut with a clean, dry cloth, elevate cut area and apply pressure until bleeding stops, then wash wound with soap and water and bandage with a sterile dressing
	Seek medical help if bleeding persists or wound is deep or gaping

SAFETY ACCESSORIES

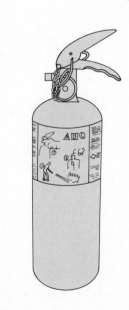

A household fire extinguisher.
Best for use in the home is a multipurpose dry-chemical extinguisher rated ABC *(left)*. A and B ratings indicate effectiveness against fires in wood or upholstery and in flammable liquids such as grease or paint, while an extinguisher rated C can be used for electrical fires. A numerical rating indicates what size fire the extinguisher can combat. An extinguisher of convenient size holds a pressurized load of 2 1/2 to 7 pounds. Check the pressure gauge monthly. After any discharge or loss of pressure, recharge or replace the tank according to the manufacturer's instructions. Mount extinguishers, using the wall brackets provided, near doors to the kitchen, utility room, garage and basement.

Two kinds of smoke alarms.
Ionization alarms *(left, top)*, which sense atomic particles, respond quickly to hot fires with little smoke, but may set off annoying false alarms in the presence of normal cooking fumes.

Photoelectric alarms *(left, bottom)* "see" smoke particles; they respond best to the smoldering typical of cooking, appliance and upholstery fires.

Install at least one smoke alarm in a central hallway, near the kitchen, bedrooms and head of the stairs, as well as in the garage and basement. Mount a battery-powered smoke alarm on the ceiling. Replace the battery once a year—the detector emits a chirping sound when the battery runs low.

CONTROLLING AN ELECTRICAL FIRE

Class ABC or BC fire extinguisher

Using a fire extinguisher. Call the fire department immediately. If there are flames or smoke coming from the walls or ceiling, leave the house to call for help. To snuff a small, accessible fire in a small appliance or at the wall outlet, use a dry-chemical fire extinguisher rated ABC or BC. Stand near an exit, 6 to 10 feet from the fire. Pull the lock pin out of the extinguisher handle and, holding the extinguisher upright, aim the nozzle at the base of the flames. Squeeze the two levers of the handle together, spraying with a quick side-to-side motion *(left)*. Keep spraying until the fire appears completely extinguished. Watch carefully for "flashback," or rekindling, and be prepared to spray again. You may also have to shut off power at the service panel to remove the source of heat or sparking causing the fire. Find the cause of the fire before using the small appliance or outlet again. Have the fire department examine the area even if the fire is out.

RESCUING A VICTIM OF ELECTRICAL SHOCK

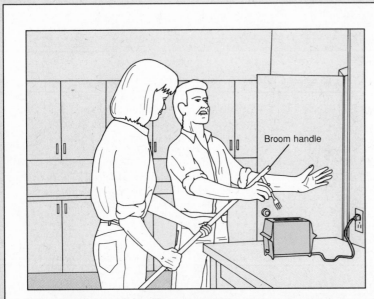

Broom handle

Freeing someone from live current. Usually, a person who contacts live current will be thrown back from the source. But muscles may contract involuntarily around a wire or appliance. If the victim is stuck, do not touch him. Disconnect the appliance by pulling its plug, or shut off power to the outlet at the main service panel. If the power cannot be cut immediately, use a wooden spoon, broom handle or chair to knock the person free *(above)*.

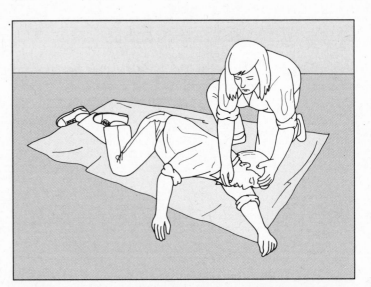

Handling a victim of electrical shock. Call for medical help immediately. Check the victim's breathing and heartbeat. If an unconscious victim is breathing and has not sustained back or neck injuries, place him in the recovery position *(above)*. Tilt the head back with the face to one side and the tongue forward to maintain an open airway. Keep the victim comfortable until help arrives. Give mouth-to-mouth resuscitation or cardiopulmonary resuscitation (CPR) only if you are qualified.

USING SMALL APPLIANCES SAFELY

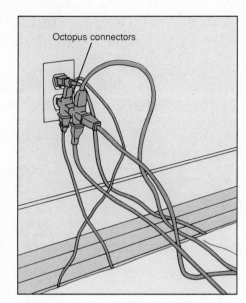

Adapters and extension cords. Do not use "octopus connectors" *(above)*. Plugging too many cords into one outlet could overload the circuit. If a plug is loose or its prongs exposed, the poor connection could produce heat and sparking. Use an extension cord only temporarily. Do not tack it down or run it under a rug.

Pulling the plug. When unplugging a small appliance, grasp the plug, not the power cord, firmly *(above)* and work the plug out of the socket. If you suspect that a plug is hot, wrap your hand in a thick towel and then pull the power cord. If you see sparks, shut off power at the main service panel first. Replace a frayed power cord immediately *(page 116)*.

Water and electricity. Wet hands can create an alternate path for electrical current. Unplug a small appliance before cleaning it with a wet cloth. Dry your hands and the appliance before plugging it back in. Never immerse an appliance in water. If it falls into water accidentally, shut off power at the service panel, then unplug and retrieve it.

Using appliances around children. Avoid leaving heating appliances such as toasters or irons unattended, especially with young children around. Place fans and heaters out of reach, or install a permanently-mounted protective screen around them *(above, left)*. Don't let power cords dangle; a small child may grab the cord and pull the appliance down on himself *(above, center)*. Kettles and irons are particular hazards, because the water inside remains hot enough to scald long after the appliance is turned off. Unplug an appliance after use and coil the power cord out of reach. Chopping and cutting appliances such as blenders and food processors, and their blades, are best stored on a high shelf *(above, right)* or in a locked cupboard when not in use.

MICROWAVE OVENS

Unlike conventional ovens, which produce heat and transfer it to the food, microwave ovens enable food to create its own heat by bombarding it with electromagnetic waves. These microwaves set food molecules vibrating, causing friction that heats the food from within.

A typical countertop microwave oven is illustrated below. Although the wiring varies somewhat from model to model, the internal components are similar. The control center of a microwave oven is either an electronic circuit board or a mechanical timer assembly. Each regulates the cooking cycle by signaling the magnetron — a microwave generator — to turn on and off at set intervals. To produce microwaves, the magnetron requires a combination of low-voltage AC current and high-voltage DC current. The transformer takes incoming household current of 120 volts AC and changes the voltage to higher and lower levels; a capacitor filters the high-voltage current and a diode converts it from AC to DC. The microwaves follow a wave guide from the magnetron to the oven cavity, and a stirrer, which is similar to a fan, distributes them evenly. With the door closed, microwaves cannot escape from the oven cavity; when it is opened, the door interlock switches turn off the oven. Several other safety switches and fuses protect both the oven and its user.

Most microwave oven problems are caused by inadequate cleaning or incorrect cooking techniques. Carefully follow the instructions for use in your owner's manual. Use a soft cloth and mild detergent to wash spattered food and grease off the oven cavity and door. If food cooks slowly or unevenly, stir or

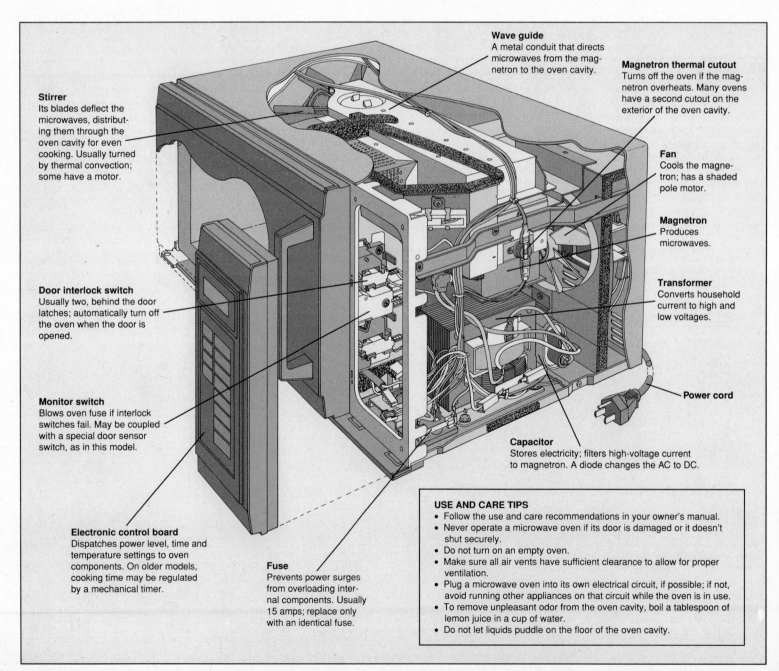

Wave guide
A metal conduit that directs microwaves from the magnetron to the oven cavity.

Magnetron thermal cutout
Turns off the oven if the magnetron overheats. Many ovens have a second cutout on the exterior of the oven cavity.

Stirrer
Its blades deflect the microwaves, distributing them through the oven cavity for even cooking. Usually turned by thermal convection; some have a motor.

Fan
Cools the magnetron; has a shaded pole motor.

Magnetron
Produces microwaves.

Door interlock switch
Usually two, behind the door latches; automatically turn off the oven when the door is opened.

Transformer
Converts household current to high and low voltages.

Monitor switch
Blows oven fuse if interlock switches fail. May be coupled with a special door sensor switch, as in this model.

Power cord

Capacitor
Stores electricity; filters high-voltage current to magnetron. A diode changes the AC to DC.

Electronic control board
Dispatches power level, time and temperature settings to oven components. On older models, cooking time may be regulated by a mechanical timer.

Fuse
Prevents power surges from overloading internal components. Usually 15 amps; replace only with an identical fuse.

USE AND CARE TIPS
- Follow the use and care recommendations in your owner's manual.
- Never operate a microwave oven if its door is damaged or it doesn't shut securely.
- Do not turn on an empty oven.
- Make sure all air vents have sufficient clearance to allow for proper ventilation.
- Plug a microwave oven into its own electrical circuit, if possible; if not, avoid running other appliances on that circuit while the oven is in use.
- To remove unpleasant odor from the oven cavity, boil a tablespoon of lemon juice in a cup of water.
- Do not let liquids puddle on the floor of the oven cavity.

turn the food halfway through the cooking cycle to balance its microwave exposure, and let it stand for a few minutes after cooking. Always put some food or drink in the oven before you turn it on; the microwaves must be absorbed or the magnetron will be damaged. Use only cookware recommended by the manufacturer, and avoid metallic containers, which can cause sparking and uneven cooking. To test whether a dish is microwave safe, place it in the oven alongside a cup of water and turn on the oven for one minute. Carefully touch the dish; if it is hot, do not use it.

In spite of its unconventional operation, a microwave oven is a sturdy, simple machine manufactured according to strict government regulations. If the oven doesn't work, first check your main service panel for a blown fuse or tripped circuit breaker. Make sure the oven is plugged into a grounded outlet; if possible, on a separate circuit from other kitchen appliances to avoid power overload.

Consult the Troubleshooting Guide below to help diagnose and fix problems, but if your oven is still under warranty, take it for professional service. Before working on any internal component, discharge the capacitor with a capacitor discharging tool *(page 114)*. Many repairs call for a multitester; to learn how to use this and other tools correctly, read Tools and Techniques *(page 110)*. After any repair, test the oven fuse *(page 16)* and make sure that no bare wires are exposed or touching the chassis. After reassembling the oven but before plugging it in, do a cold check for leaking voltage *(page 114)*, then take the oven for a microwave leakage test.

TROUBLESHOOTING GUIDE

continued ►

SYMPTOM	POSSIBLE CAUSE	PROCEDURE
Microwave oven doesn't work at all	Oven unplugged or turned off, or door not closed properly	Plug in oven, close door securely and turn on oven
	No power to outlet or outlet faulty	Reset breaker or replace fuse *(p. 112)* □○; have outlet serviced
	Power cord faulty	Test and replace power cord *(p. 116)* ◨○▲
	Oven fuse blown	Test and replace oven fuse *(p. 16)* ◨○
	Door interlock switch or monitor switch faulty	Check oven fuse *(p. 16)* ◨○; test and replace interlock switches *(p. 16)* □○ and monitor switch *(p. 17)* ◨○
	Thermal cutout faulty	Test and replace thermal cutouts *(p. 17)* ◨○
	Circuit board faulty	Take oven for professional service
Oven doesn't cook, but display lights are on	Door interlock switch or monitor switch faulty	Test and replace interlock switches *(p. 16)* □○ and monitor switch *(p. 17)* ◨○
	Thermal cutout faulty	Test and replace thermal cutouts *(p. 17)* ◨○
	Door, triac or circuit board faulty	Take oven for professional service
Oven doesn't cook, but emits humming sound	Transformer faulty	Test and replace transformer *(p. 18)* ■○▲
	Capacitor faulty	Test and replace capacitor *(p. 19)* ◨○▲
	Diode faulty	Test and replace diode *(p. 19)* ◨○▲
	Magnetron faulty	Take oven for professional service
Oven starts cooking, then stops	Other appliances on circuit draining power	Reset breaker or replace fuse *(p. 112)* □○ and plug oven into separate circuit
	Power cord faulty	Check oven fuse *(p. 16)* ◨○; test power cord *(p. 116)* ◨○▲
	Door interlock switch or monitor switch faulty	Check oven fuse *(p. 16)* ◨○; test and replace interlock switches *(p. 16)* □○ and monitor switch *(p. 17)* ◨○
	Thermal cutout faulty	Test and replace thermal cutouts *(p. 17)* ◨○; check fuse *(p. 16)* ◨○
	Fan jammed or fan motor faulty	Service fan motor *(p. 20)* ◨◕▲; check oven fuse *(p. 16)* ◨○
	Transformer faulty	Test and replace transformer *(p. 18)* ■○▲; check fuse *(p. 16)* ◨○
	Capacitor faulty	Test and replace capacitor *(p. 19)* ◨○▲; check fuse *(p. 16)* ◨○
	Diode faulty	Test and replace diode *(p. 19)* ◨○▲; check oven fuse *(p. 16)* ◨○
	Circuit board, triac or magnetron faulty	Take oven for professional service
Oven cooks slowly or intermittently	Power cord faulty	Test and replace power cord *(p. 116)* ◨○▲
	Voltage too low at wall outlet	Call an electrician for service
	Door interlock switch or monitor switch faulty	Check oven fuse *(p. 16)* ◨○; test and replace interlock switches *(p. 16)* □○ and monitor switch *(p. 17)* ◨○
	Fan jammed or fan motor faulty	Check whether blades turn freely; service fan motor *(p. 20)* ◨◕▲

DEGREE OF DIFFICULTY: □ **Easy** ◨ **Moderate** ■ **Complex**
ESTIMATED TIME: ○ **Less than 1 hour** ◕ **1 to 3 hours** ● **Over 3 hours**   ▲ **Special tool required**

TROUBLESHOOTING GUIDE (continued)

SYMPTOM	POSSIBLE CAUSE	PROCEDURE
Oven cooks slowly or intermittently (continued)	Thermal cutout faulty	Test and replace thermal cutouts *(p. 17)* ◨○
	Mechanical timer assembly faulty	Test and replace mechanical timer assembly *(p. 21)* ◨○▲
	Transformer wire connections loose	Check wire connections; repair if necessary *(p. 118)* □○
	Circuit board, triac or magnetron faulty	Take oven for professional service
Oven burns or undercooks food regularly	Timer programmed incorrectly	Consult owner's manual and manufacturer's cookbook
	Mechanical timer assembly faulty	Test and replace mechanical timer assembly *(p. 21)* ◨○▲
	Stirrer, circuit board or triac faulty	Take oven for professional service
Oven cooks food unevenly	Cooking methods incorrect	Consult manufacturer's cookbook
	Dish not microwave safe	Consult owner's manual and use correct dish
	Microwave distribution uneven	Rearrange food halfway through cycle; install windup turntable
	Stirrer dirty or faulty, or oven cavity lining chipped or rusted	Take oven for professional service
Oven doesn't turn off	Mechanical timer assembly faulty	Test and replace mechanical timer assembly *(p. 21)* ◨○▲
	Circuit board or triac faulty	Take oven for professional service
Temperature probe doesn't work	Controls not set for probe, or probe faulty	Check setting for probe; test and replace probe *(p. 14)* ◨○▲
	Probe jack greasy or clogged	Unplug oven; clean jack with foam swab dipped in soapy water
	Probe jack or circuit board faulty	Take oven for professional service
Door doesn't close properly	Door misaligned or gasket damaged	Take oven for professional service
Oven light doesn't work	Bulb burned out	Consult owner's manual and replace bulb
	Thermal cutout faulty	Test and replace thermal cutouts *(p. 17)* ◨○
Oven light doesn't turn off	Circuit board faulty	Take oven for professional service
Sparks in oven cavity while cooking	Dish, metal food wrap not microwave safe	Consult owner's manual and use correct dish or wrap
	Oven cavity lining chipped or rusted, or magnetron faulty	Take oven for professional service
Oven gives electrical shock	Plug's ground prong missing	Replace power cord *(p. 116)* ◨○▲
	Components not properly grounded	Take oven for professional service

DEGREE OF DIFFICULTY: □ Easy ◨ Moderate ■ Complex
ESTIMATED TIME: ○ Less than 1 hour ◑ 1 to 3 hours ● Over 3 hours ▲ Special tool required

SERVICING THE TEMPERATURE PROBE

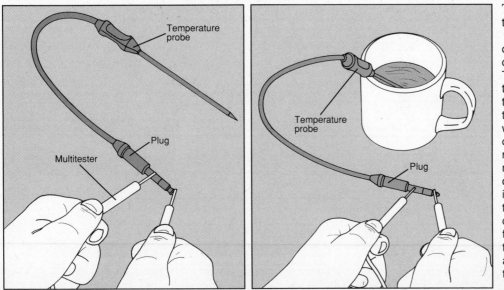

Testing the temperature probe. Set a multi-tester at RX10K to test resistance *(page 113)*. Touch one multitester probe to the tip of the plug and the second probe to the shaft of the plug, above the black ring *(far left)*. The multitester should register partial resistance. If the plug has infinite or zero resistance, replace the temperature probe. If it tests OK, perform a heat test: Heat a cup of water to boiling. Insert the thermometer end of the temperature probe into the water and repeat the resistance test *(near left)*. The multitester needle should swing to the middle of the scale, then gradually fall back toward infinite resistance as the water cools. If the temperature probe tests OK, suspect a faulty circuit board and take the microwave oven for professional service. If the probe fails any test, replace it with an exact duplicate from a microwave oven service center or from the manufacturer.

ACCESS TO OVEN COMPONENTS

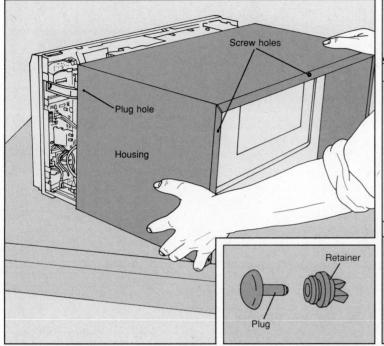

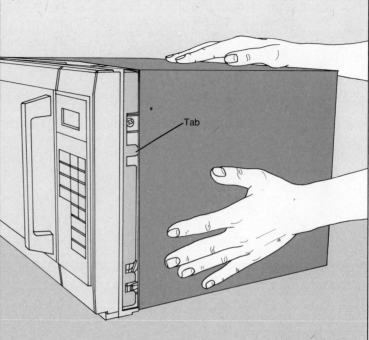

Removing and replacing the housing. Before working on a microwave oven, make a capacitor discharging tool *(page 114)*. If the oven is mounted under a cabinet or on the wall, consult the manufacturer's installation guide for the correct dismounting procedures. Unplug a countertop microwave oven and place it on a sturdy work table. The metal housing on the model shown above is secured to the oven chassis by screws in the back and by plastic plug fasteners *(inset)* through the sides. Use a screwdriver to remove all screws. Extract the plugs from their retainers using a narrow-tipped screwdriver padded with masking tape. Some oven housings are secured by screws with splined washers; keep the washers for replacement later. When all screws and plugs have been removed, slide off the oven housing *(above, left)*. You now have access to the oven's internal components. Before servicing any of them, use the discharging tool to discharge the capacitor *(step below)*. After working on the microwave oven, reinstall the housing. Adjust its position so that the metal tabs on the front of the oven chassis fit into the slots on the housing *(above, right)*. Reinstall all screws and plastic plugs and cold check for leaking voltage *(page 114)* before plugging in the oven.

DISCHARGING THE CAPACITOR

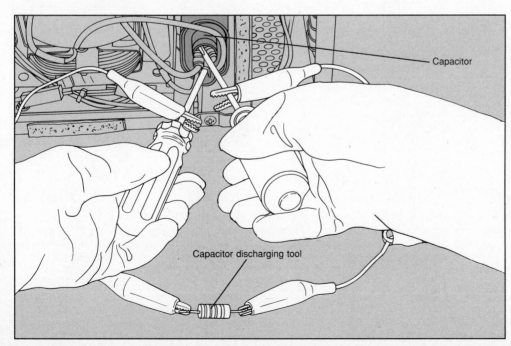

Discharging a capacitor. Caution: Capacitors store potentially dangerous voltage. Unplug the oven and remove its housing *(step above)*. Before working on any internal components, wait five minutes and then use a capacitor discharging tool *(page 114)* to drain the capacitor of stored charge. Locate the capacitor, an oblong metal canister with two terminals. Without touching any internal component, touch one screwdriver blade of the discharging tool to one capacitor terminal and the other blade to the other terminal at the same time, for one second *(left)*. The capacitor may spark as it discharges.

TESTING AND REPLACING THE OVEN FUSE

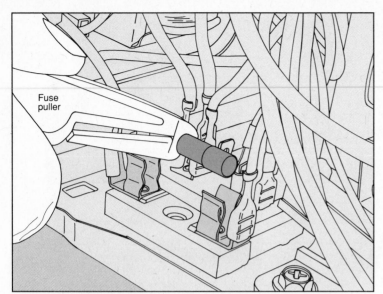

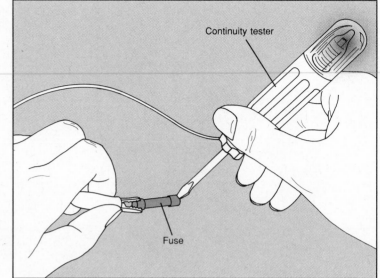

1 **Removing the fuse.** Unplug the oven, remove the housing and discharge the capacitor *(page 15)*. On the model shown here, the fuse is clearly visible on a junction board on the chassis floor. On some models, the fuse may be hidden by oven components; locate it by tracing the black wire and white wire from the power cord to their terminals on the power block next to the fuse. To remove the fuse for testing, grip it with a fuse puller and gently pry it out of its retaining clips *(above)*.

2 **Testing the fuse.** To test the oven fuse for continuity, use a continuity tester *(page 113)*. Touch a probe to each end of the fuse *(above)*. If the tester lights, reinstall the fuse. If it doesn't, replace the fuse with an exact duplicate purchased at an electronics parts supplier. Gently push the fuse into its retaining clips and reinstall the oven housing *(page 15)*. Plug in the oven. If the fuse blows repeatedly, consult the Troubleshooting Guide *(page 13)* for other possible causes for the problem.

SERVICING A DOOR INTERLOCK SWITCH

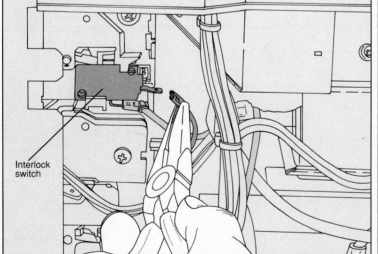

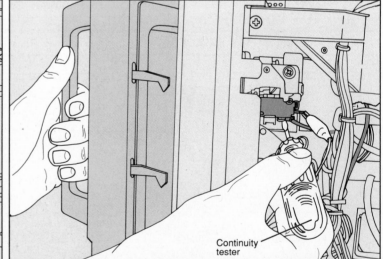

Testing an interlock switch. Unplug the oven, remove the housing and discharge the capacitor *(page 15)*. Locate the two door interlock switches behind the door latches. If a wire is loose or broken, repair the wire connection *(page 118)*. To test an interlock switch, use long-nose pliers to disconnect one wire from its terminal *(above, left)*. Attach the alligator clip of a continuity tester to this terminal and touch the probe to the second terminal *(above, right)*. With the oven door closed, the tester should light, and with the door open, the tester should not light. Test the second interlock switch the same way. If both switches test OK, return to the Trou-

bleshooting Guide *(page 13)* to check for other possible causes of oven malfunction before reinstalling the housing *(page 15)*. If either switch fails this test, replace it: Label and disconnect the second wire. Unscrew or unclip the faulty switch from its mounting block and buy an exact replacement from a microwave oven service center or the manufacturer. Install the new switch, reversing the steps you took for removal. Reinstall the housing *(page 15)* and take the oven to a microwave oven service center for a microwave leakage test. If you have trouble removing a switch, reinstall the housing *(page 15)* and take the oven for professional service.

SERVICING THE MONITOR SWITCH

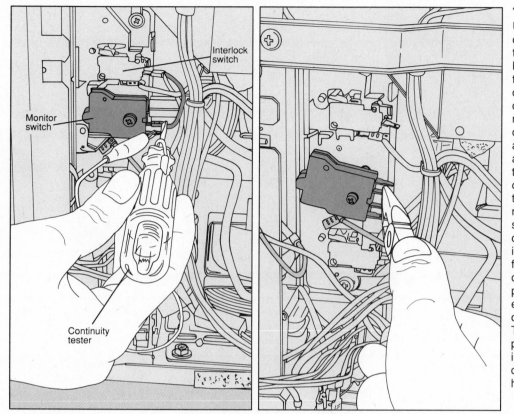

Monitor switch

Interlock switch

Continuity tester

Testing and replacing the monitor switch.
Unplug the oven, remove the housing and discharge the capacitor *(page 15)*. Locate the monitor switch, between the door interlock switches behind the door latches. In the model shown here, it is coupled with a door sensor switch. Use long-nose pliers to disconnect the wire from the lowest terminal on the switch. Attach the alligator clip of a continuity tester to the free terminal and touch the probe to the terminal directly above it *(far left)*. With the door open, the tester should light, and with the door closed, it should not light. If the switch fails this test, replace it. If the switch tests OK, reconnect the wire and consult the Troubleshooting Guide *(page 13)* for other possible causes of oven malfunction before reinstalling the housing *(page 15)*. To replace a faulty switch, label and disconnect its second wire. Unscrew or unclip the switch and pull it off its mounting plate. Purchase an exact replacement switch from a microwave oven service center or the manufacturer. To install a new switch, use long-nose pliers to position the switch on the mounting plate *(near left)*, then screw or clip it on. Reconnect the wires and reinstall the housing *(page 15)*.

TESTING AND REPLACING A THERMAL CUTOUT

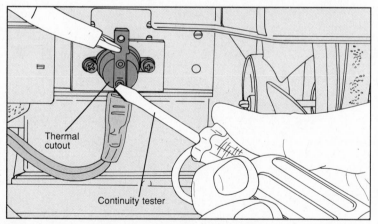

Thermal cutout

Continuity tester

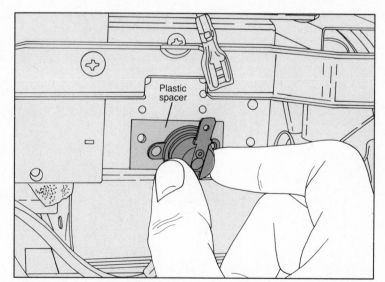

Plastic spacer

1 **Testing the thermal cutouts.** Unplug the oven, remove the housing and discharge the capacitor *(page 15)*. Locate the round, flat thermal cutout mounted on the magnetron. To test the cutout for continuity, disconnect one of its wires and attach the alligator clip of a continuity tester to the free terminal. Touch the probe to the second terminal *(above)*. The tester should light. If it doesn't, go to step 2 to replace the cutout. If it tests OK, reconnect the wire. Many microwave ovens have a second thermal cutout on the other side of the oven or on top. Test it the same way. If both thermal cutouts test OK, consult the Troubleshooting Guide *(page 13)* for other possible causes of oven malfunction before reinstalling the housing *(page 15)*.

2 **Replacing a thermal cutout.** Label and disconnect the second wire. Use a screwdriver to remove the two screws holding the cutout in place and take it off, noting its exact position. If there is a plastic spacer on the magnetron's thermal cutout, keep it for reuse. Buy an identical thermal cutout from a microwave oven service center or the manufacturer. Mount the new cutout in the same spot, using the screws and spacer you removed. Reconnect the wires and reinstall the housing *(page 15)*.

SERVICING THE TRANSFORMER

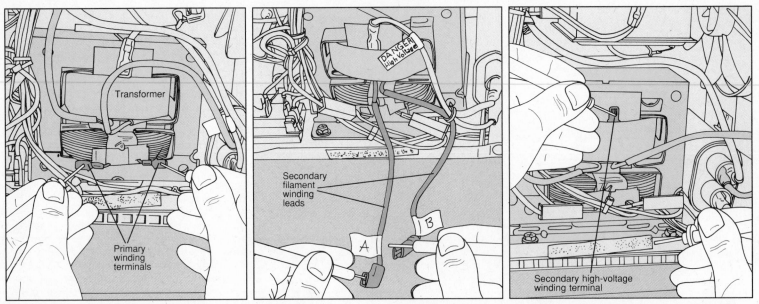

1 **Testing the transformer.** Unplug the oven, remove the housing and discharge the capacitor *(page 15)*. Locate the transformer, a large metal block with exposed windings. If the transformer smells or looks burned, take the oven for professional service. To test each of the three transformer windings for less obvious damage, set a multitester to RX1 *(page 113)*. First locate the primary winding, typically visible through an opening in the transformer. Label and disconnect the winding's two wires and touch a probe to each terminal *(above, left)*. The multitester should indicate very low resistance. If not, replace the transformer *(step 2)*. If the winding tests OK, reconnect the wires and test the secondary filament winding. Identify this winding by its two insulated wires that lead to terminals on the magnetron and the capacitor. Label and

disconnect the wires. Touch a probe to each wire *(above, center)*. The multitester should indicate very low resistance. If not, replace the transformer *(step 2)*. If it tests OK, reconnect the wires and go on to test the secondary high-voltage winding, which is generally hidden behind the manufacturer's label. If the transformer winding has only one terminal as shown above, disconnect the wire and test between the terminal and the chassis *(above, right)*. If the winding has two terminals, label and disconnect both wires and test between both terminals. The multitester should indicate some resistance. If not, replace the transformer *(step 2)*. If the windings test OK, suspect a faulty triac or control board and take the oven for professional service.

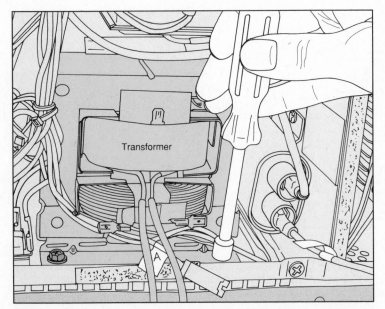

2 **Removing the transformer.** Label each transformer wire lead by its winding and position, and use long-nose pliers to disconnect it from its terminal. Locate the screws or bolts that secure the transformer to the microwave oven and unscrew them *(above)*. Pull the transformer out of the oven.

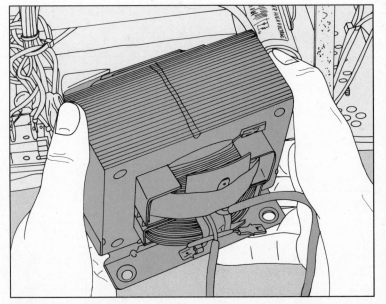

3 **Installing a new transformer.** Purchase an identical replacement transformer from a microwave oven service center or from the manufacturer. Position the new transformer *(above)*, securing it tightly to the chassis floor using the same screws or bolts you removed. Reconnect all wires to their correct terminals, and reinstall the oven housing *(page 15)*.

SERVICING THE CAPACITOR

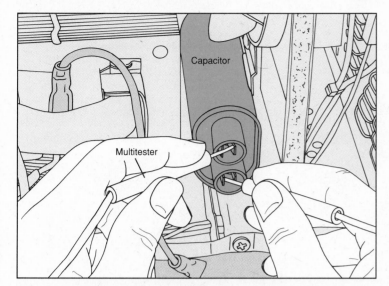

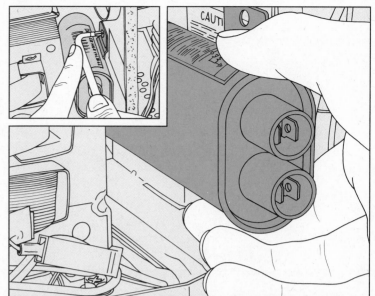

1 **Testing the capacitor.** Unplug the oven, remove the housing and discharge the capacitor *(page 15)*. Inspect the capacitor and replace one that is bulging or cracked *(step 2)*. To check for less evident damage, set a multitester to RX10K *(page 113)*. Label the wires and disconnect them from the capacitor terminals. Touch a tester probe to each terminal *(above)*. The needle should rise, then fall immediately. If the capacitor tests OK, return to the Troubleshooting Guide *(page 13)* and check for other possible causes for oven malfunction before reinstalling its housing *(page 15)*. If the multitester needle swings to the right and stays, or if it doesn't move at all, replace the capacitor *(step 2)*.

2 **Replacing the capacitor.** Locate and remove the screws securing the capacitor to the oven chassis. On the model shown here, the capacitor mounting screw is difficult to reach; use an offset screwdriver to remove it *(inset)* and pull the capacitor free *(above)*. It may be necessary to remove the transformer *(page 18)* and/or the fan blade *(page 20)* before lifting out the capacitor. Purchase an exact replacement capacitor from a microwave oven service center or the manufacturer and screw it in place. Reconnect all wires and reinstall the housing *(page 15)*.

TESTING THE DIODE

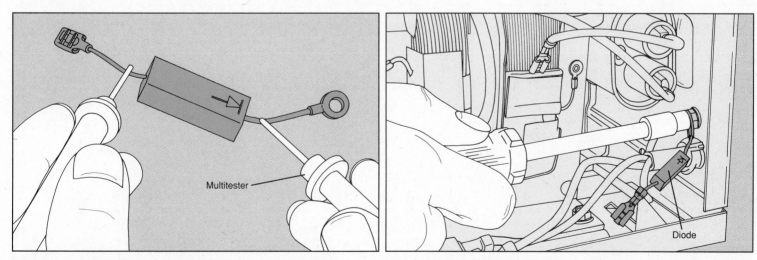

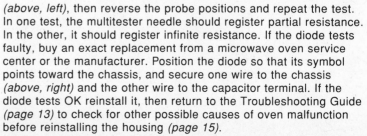

Testing the diode. Unplug the oven, remove the housing and discharge the capacitor *(page 15)*. Locate the diode, a small component connected to one of the capacitor terminals and to the chassis. If the diode isn't visible, it is encased within the capacitor housing and is difficult to test; take the oven for professional service. To test an externally-mounted diode, use a multitester set to its highest ohms setting to test resistance *(page 113)*. Pull the diode connector off the capacitor terminal with long-nose pliers and unscrew the diode's ground wire from the chassis using a nut driver or screwdriver. Touch a multitester probe to each wire end *(above, left)*, then reverse the probe positions and repeat the test. In one test, the multitester needle should register partial resistance. In the other, it should register infinite resistance. If the diode tests faulty, buy an exact replacement from a microwave oven service center or the manufacturer. Position the diode so that its symbol points toward the chassis, and secure one wire to the chassis *(above, right)* and the other wire to the capacitor terminal. If the diode tests OK reinstall it, then return to the Troubleshooting Guide *(page 13)* to check for other possible causes of oven malfunction before reinstalling the housing *(page 15)*.

INSPECTING AND REPLACING THE FAN MOTOR

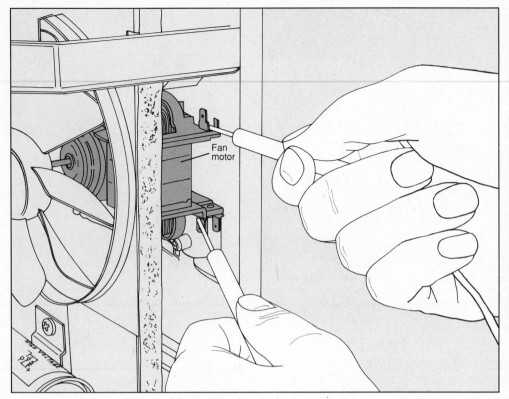

1 **Testing the fan motor.** To check whether the fan is operating, first remove the oven housing *(page 15)*. **Caution:** Do not touch any internal components. Put a cup of water in the oven cavity, close the door, plug in the oven and turn it on for less than a minute. Do not touch the microwave oven. Observe the fan; if it does not revolve, unplug the oven, wait five minutes and discharge the capacitor *(page 15)*. Remove any obstructions that may be preventing the fan from rotating. Turn the blades by hand; if they don't move easily, disconnect the wires from the motor and take out the motor *(steps 2-5)* for cleaning and lubrication *(page 125)*. If the fan blades move freely, check for a burning smell and examine the motor windings for discoloration. If a wire is loose or broken, repair the wire connection *(page 118)*. To test the motor for less visible damage, set a multitester to test resistance *(page 113)*. Label and disconnect the wires. Touch a multitester probe to each motor terminal *(left)*. The motor should have low resistance. If not, replace it. If the motor tests OK, return to the Troubleshooting Guide *(page 13)* and check for other causes of oven malfunction before reinstalling its housing *(page 15)*.

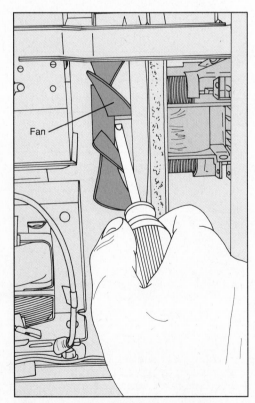

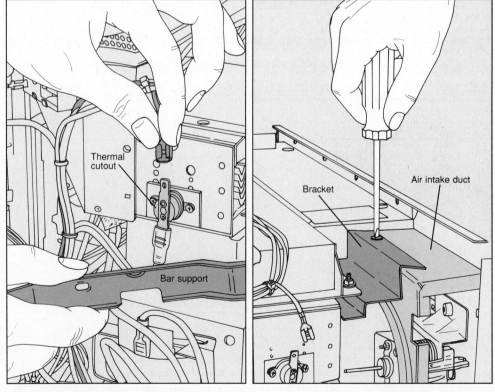

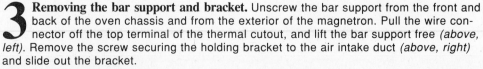

2 **Removing the fan.** Fan motors are mounted in various ways. On this model, remove several parts to access the motor: First, slip a screwdriver between the fan blades and the intake duct and pry the fan off the motor shaft *(above)*.

3 **Removing the bar support and bracket.** Unscrew the bar support from the front and back of the oven chassis and from the exterior of the magnetron. Pull the wire connector off the top terminal of the thermal cutout, and lift the bar support free *(above, left)*. Remove the screw securing the holding bracket to the air intake duct *(above, right)* and slide out the bracket.

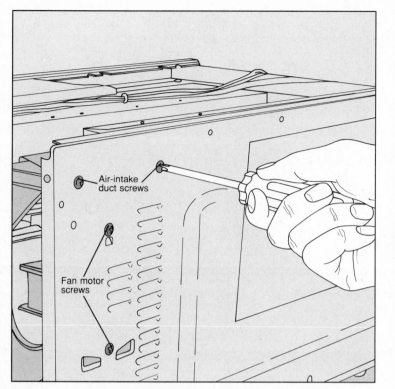

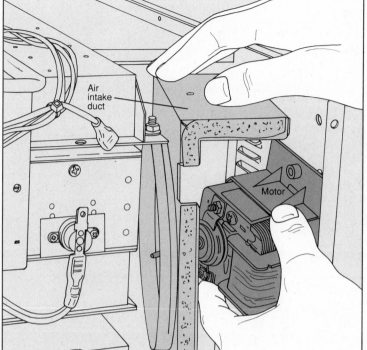

4 **Freeing the fan motor.** Remove the screws securing the air intake duct to the chassis *(above)*; on this model, two screws attach the duct to the back, while one screw attaches it to the floor. Then unscrew the motor from the back of the chassis. Do not remove more screws than necessary. Push the air intake duct toward the magnetron. If the duct is secured to the exterior of the oven cavity with a clip, lift the duct over the clip and then push the duct aside.

5 **Replacing the fan motor.** The motor is held in a casing that is in turn slotted into the chassis. Firmly grasp the motor and slide the casing out of the slots *(above)*. Then angle the motor shaft inward and pull the motor free. Service the motor *(page 124)* or purchase an exact replacement from a microwave oven service center or the manufacturer. To install the new motor, reverse the steps taken for removal, and reinstall the housing *(page 15)*.

SERVICING A MECHANICAL TIMER ASSEMBLY

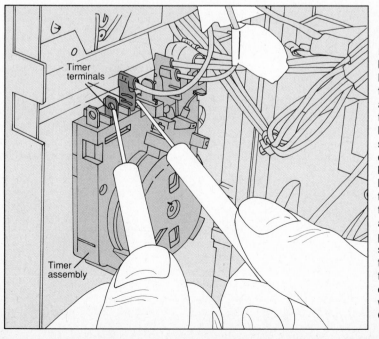

1 **Testing the timer switch.** Set the timer knob to 1 minute, place a cup of water in the oven cavity, and turn on the oven. Check the timer's accuracy with a wristwatch. If the timer keeps the correct time, suspect a faulty circuit board and take the oven for professional service. If the timer loses or gains time, replace the timer assembly *(step 3)*. If the timer knob doesn't move at all, turn on the oven again and press your fingers against the control panel next to the timer knob. If you do not feel a vibration, the timer motor may be faulty; test it *(step 2)*. If you feel a vibration, test the timer switch: Turn off the microwave, remove the housing and discharge the capacitor *(page 15)*. Locate the timer assembly behind the timer knob. If a wire connection is broken or loose, repair it *(page 118)*. Turn on the timer and set a multitester to test resistance, usually RX100 *(page 113)*. Some timer switches, like the one here, have three terminals. Label and disconnect two of the wires and test all three possible pairings of terminals *(left)*. The multitester should indicate some resistance at least once. If the timer switch has only two terminals, pull off one wire and test for continuity between the terminals *(left)*. The multitester should indicate continuity. If the switch doesn't show resistance or continuity when it should, replace the timer assembly *(step 3)*. If it tests OK, test the timer motor *(step 2)*.

SERVICING A MECHANICAL TIMER ASSEMBLY (continued)

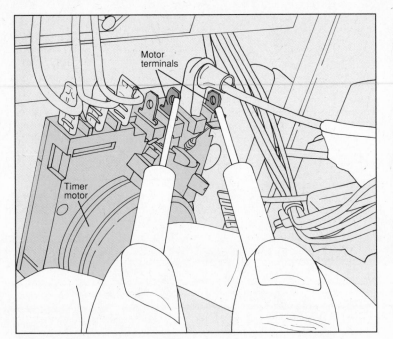

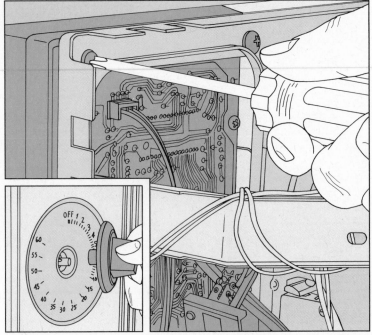

2 **Testing the timer motor.** Unplug the oven, remove the housing and discharge the capacitor *(page 15)*. The timer motor has two terminals, one wired to the fan motor and a thermal cutout, and a second wired to the circuit board. Label and disconnect a wire from each terminal. Set a multitester to RX100 *(page 113)* and touch a probe to each terminal *(above)*. The multitester should indicate partial resistance. If the timer motor tests faulty, replace the timer assembly *(step 3)*. If it tests OK, suspect a faulty circuit board and take the oven for professional service.

3 **Removing the control panel frame.** With the housing removed and the capacitor discharged *(page 15)*, pull all knobs and buttons straight off the control panel *(inset)*. If a knob won't budge, slip a cloth behind it and pull the cloth to dislodge it. Label all wires on the back of the control panel, and disconnect them. Remove the screws attaching the control panel frame to the oven chassis *(above)*. Lift the panel frame off the front of the oven and set it face down on a padded work surface.

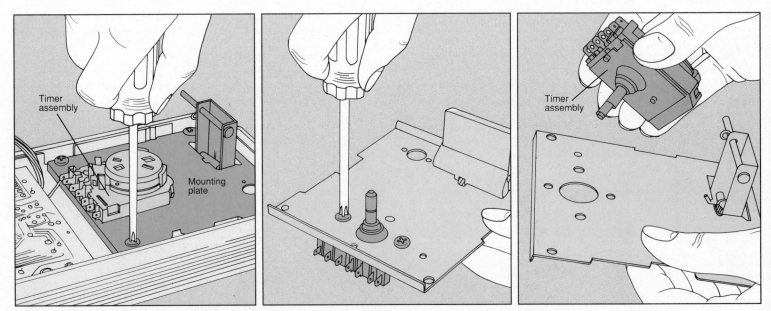

4 **Replacing the timer assembly.** Remove the screws attaching the timer-assembly mounting plate to the control panel frame *(above, left)*, then lift off the mounting plate. Turn the plate over, placing the timer assembly shaft up on the work surface. Unscrew the two screws securing the timer assembly to the mounting plate *(above, center)* and pull it off the plate *(above, right)*. Purchase an identical replacement timer assembly from a microwave oven service center or from the manufacturer. To install a new timer, reverse the steps taken for removal. Reattach the control panel frame to the oven chassis and reconnect all wires to the circuit board and the timer assembly. Reinstall the housing *(page 15)*.

TOASTERS AND TOASTER OVENS

The ubiquitous toaster and its versatile cousin, the toaster oven, are among the most frequently used of small kitchen appliances. Even so, with regular cleaning and proper care they will operate for years before breaking down. Consult the Troubleshooting Guide below when a problem does arise. While cheaper models are uneconomical to repair, more expensive toasters and toaster ovens can be fixed at a fraction of their replacement price. Keep in mind, however, that some toaster parts are now riveted to the chassis, and most toaster-oven connections are spot welded. Replace rivets as described in this chapter *(page 26)*. Replace a welded connection with silver solder; consult Tools & Techniques *(page 121)*.

The deceptive ease with which a toaster browns bread belies a complicated interaction of mechanical and electrical events.

When the carriage lowers, it catches a latch that closes a main switch, completing a circuit to the heating elements. In some toasters a thermostat alone controls the heat, but in the model shown on page 24, the thermostat works with a solenoid switch and solenoid. When the thermostat's metal arm bends with the heat, it closes the switch, allowing current to magnetize the solenoid. The solenoid then pulls a metal latch release, freeing the latch so that the carriage can pop up.

The toaster oven works on virtually the same principles as a toaster, but since it also bakes and broils it has a radically different design. As illustrated on page 27, a typical toaster oven has two thermostats, one for toasting, the other for baking. In the unlikely event that either breaks, their repair is best left to an authorized repair center.

TROUBLESHOOTING GUIDE

SYMPTOM	POSSIBLE CAUSE	PROCEDURE
TOASTERS		
Carriage lowers and latches, but toaster doesn't turn on	No power to outlet or outlet faulty	Reset breaker or replace fuse *(p. 112)* □○; have outlet serviced
	Power cord faulty	Test and replace power cord *(p. 116)* ◪○▲
	Main switch contacts broken	Access chassis; replace if necessary *(p. 25)* ◪○; replace toaster
Carriage lowers or latches stiffly or not at all	Slide rod or latches dirty or obstructed	Service carriage-and-latch assembly *(p. 26)* ◪○
Toaster buzzes when latched and carriage pops up instantly	Toaster not cooling after toasting	Select darker color setting or wait longer between batches of toast
	Solenoid switch faulty	Test and replace solenoid switch *(p. 25)* ◪○
Toaster stays on after carriage pops up	Main switch contacts fused	Access chassis and replace chassis if necessary *(p. 25)* ◪○; or replace toaster
Carriage doesn't pop up and toast burns	Thermostat calibrated incorrectly	Recalibrate thermostat *(p. 24)* □○
	Solenoid switch or solenoid faulty	Test, replace solenoid switch *(p. 25)* ◪○ or solenoid *(p. 26)* ◪○▲
	Thermostat faulty	Replace toaster
Toast too dark or too light	Thermostat calibrated incorrectly	Recalibrate thermostat *(p. 24)* □○
Bread toasts on only one side	Element or elements faulty	Replace chassis *(p. 25)* ◪○; or replace toaster
Toaster gives electrical shock	Internal component grounded to chassis	Take toaster for professional service
TOASTER OVENS		
Toaster oven doesn't work on toaster function and oven function	No power to outlet or outlet faulty	Reset breaker or replace fuse *(p. 112)* □○; have outlet serviced
	Power cord faulty	Test and replace power cord *(p. 116)* ◪○▲
	Main switch faulty	Service main switch *(p. 28)* ◪○
	Thermal fuse faulty	Test and replace thermal fuse *(p. 30)* ◪○
Toaster oven works on only one function	Oven control thermostat or toaster switch faulty	Take toaster oven for professional service
Toaster function doesn't turn off unless door is opened	Solenoid faulty	Test and replace solenoid *(p. 29)* ◪○▲
	Toaster switch or toaster thermostat faulty	Take toaster oven for professional service
Toaster function buzzes and latch doesn't stay down	Toaster thermostat faulty	Take toaster oven for professional service
Toaster oven stays on when door is opened	Main switch faulty	Service main switch *(p. 28)* ◪○
Upper or lower elements don't heat	Upper or lower element faulty	Test and replace element *(p. 30)* ◪○▲
	Oven or toaster thermostat faulty	Take toaster oven for professional service
Oven gives electrical shock	Internal component grounded to chassis	Take toaster oven for professional service

DEGREE OF DIFFICULTY: □ Easy ◪ Moderate ■ Complex
ESTIMATED TIME: ○ Less than 1 hour ◑ 1 to 3 hours ● Over 3 hours

▲ Special tool required

TOASTERS

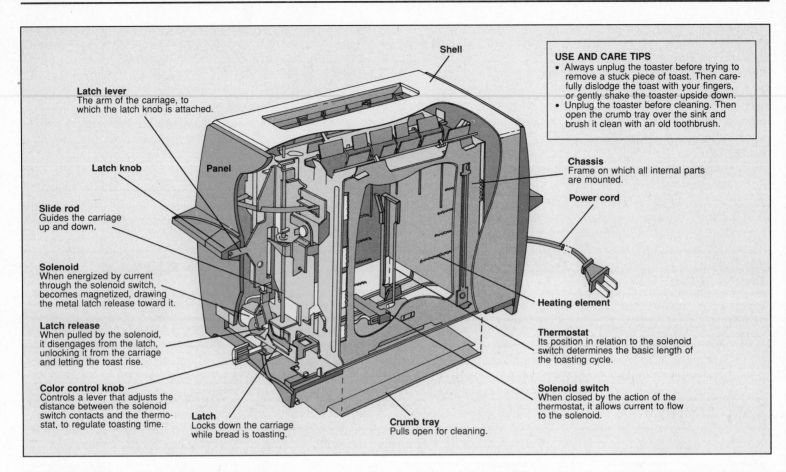

Shell

USE AND CARE TIPS
- Always unplug the toaster before trying to remove a stuck piece of toast. Then carefully dislodge the toast with your fingers, or gently shake the toaster upside down.
- Unplug the toaster before cleaning. Then open the crumb tray over the sink and brush it clean with an old toothbrush.

Latch lever
The arm of the carriage, to which the latch knob is attached.

Latch knob

Panel

Slide rod
Guides the carriage up and down.

Solenoid
When energized by current through the solenoid switch, becomes magnetized, drawing the metal latch release toward it.

Latch release
When pulled by the solenoid, it disengages from the latch, unlocking it from the carriage and letting the toast rise.

Color control knob
Controls a lever that adjusts the distance between the solenoid switch contacts and the thermostat, to regulate toasting time.

Latch
Locks down the carriage while bread is toasting.

Crumb tray
Pulls open for cleaning.

Chassis
Frame on which all internal parts are mounted.

Power cord

Heating element

Thermostat
Its position in relation to the solenoid switch determines the basic length of the toasting cycle.

Solenoid switch
When closed by the action of the thermostat, it allows current to flow to the solenoid.

RECALIBRATING THE THERMOSTAT

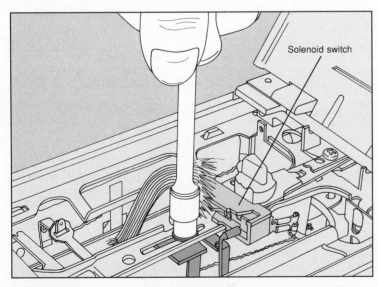

Solenoid switch

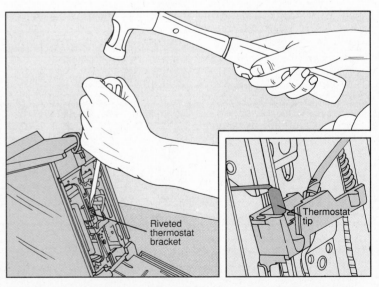

Riveted thermostat bracket

Thermostat tip

Adjusting the thermostat. Unplug the toaster and let it cool. Set the color control knob to its middle position. Turn over the toaster and open the crumb tray. Locate the thermostat bracket in a slot in the chassis. It may be screwed or riveted to the thermostat. Move the bracket toward the solenoid switch for a shorter toasting cycle and lighter-colored toast. Move the bracket away from the solenoid switch for a longer toasting cycle and darker-colored toast. To move a screwed bracket, loosen the screw *(above, left)* and slide the bracket with your fingers. Move a riveted bracket by placing a screwdriver blade at its edge and very gently tapping it with a hammer *(above, right)*. Never move the ceramic thermostat tip closer than 3/16 inch from the outer solenoid-switch contact *(inset)*. After moving a screwed bracket, retighten the screw. Close the crumb tray. Turn over and plug in the toaster and run a test cycle of toast. If the toast is still not the right color, repeat the adjustment until you get the desired result.

DISASSEMBLING AND INSPECTING A TOASTER

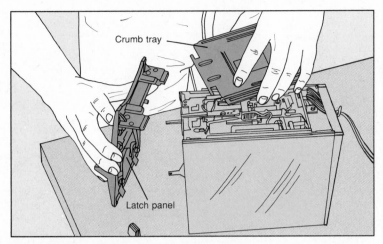

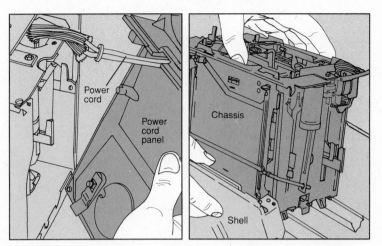

Removing the crumb tray and latch panel. Unplug the toaster and let it cool. Turn it over and open the crumb tray. Remove the color control knob; either unscrew the knob and pull it and its lever out of the side of the toaster, or just pull the knob off. Then remove the latch knob; slip the tips of a pair of long-nose pliers behind it on each side of its lever and pry it off, taking care not to lose the tiny clip that fits through a hole in the end of the lever to hold the knob in place. Unscrew the bottom of the latch panel from the chassis. Pull out on the bottom of the panel and lift off the crumb tray. Pull the latch panel up and out over the ends of the latch lever and color control lever, if it's still in place *(above)*. You can now service the latch assembly *(page 26)*.

Accessing and removing the chassis. To inspect the main switch contacts, remove the power cord panel: Unscrew the bottom of the panel and pull it away from the toaster. Then pull up on the panel and slide it along the power cord *(above, left)*. Check the main switch contacts; if they look fused or broken, replace the chassis or the toaster. To remove the chassis for replacement or to service the solenoid, pull off the crumb tray and latch panel *(left)*. Pull the chassis out of the shell by bending first one side of the shell and then the other, working its corner tabs free of the chassis *(above, right)*. If installing a new chassis, buy an exact replacement from an authorized service center and install it by reversing the steps you took to remove the old one. Cold check for leaking voltage *(page 114)* before plugging in the toaster.

SERVICING THE SOLENOID SWITCH

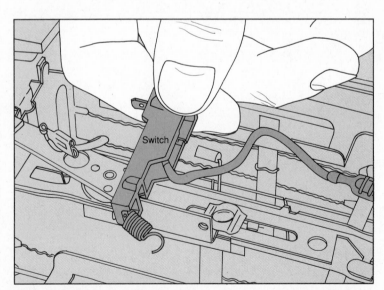

1 **Testing the solenoid switch.** Unplug the toaster and let it cool. Turn it over and open the crumb tray. The solenoid switch is attached to an element connecting rod and to a wire from the solenoid. Pull the switch wire connector off the rod and pull the solenoid wire connector off the switch. Clip a continuity tester to the switch wire connector and touch the tester probe to the switch terminal. The tester should not light. Press the ceramic thermostat tip firmly against the solenoid switch contacts *(above)*. The tester should light. If the solenoid switch fails either test, go to step 2 to replace the switch.

2 **Replacing the solenoid switch.** Use long-nose pliers to twist the solenoid switch spring out of its hole in the chassis, and take out the switch *(above)*. Unhook the spring from the switch and keep it for reuse. Buy an exact replacement switch from an authorized service center. Hook the spring onto the chassis, place the switch in position and use long-nose pliers to rehook the spring. Attach the switch wire connector to the element connecting-rod terminal and attach the solenoid wire connector to the switch terminal. Reassemble the toaster, reversing the steps taken to disassemble it, and cold check for leaking voltage *(page 114)*.

SERVICING THE CARRIAGE-AND-LATCH ASSEMBLY

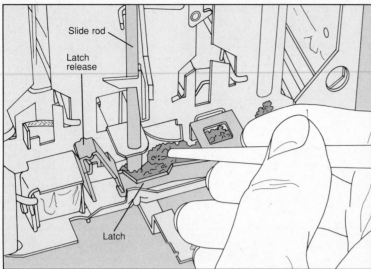

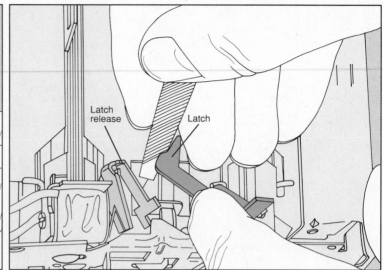

Inspecting and cleaning the carriage-and-latch assembly. Remove the crumb tray and latch panel *(page 25)*. Use a a tool such as a wooden stick to clean away crumbs and deposits from the chassis floor *(above, left)*. Turn the toaster on its side and shake it to loosen hard-to-reach pieces of toast or foreign objects. Lower and raise the latch lever along the slide rod a few times; if the lever moves stiffly, lubricate the rod. Slip a piece of paper between the rod and the chassis and spray the rod lightly with a petroleum- based lubricant. Do not get lubricant on the chassis or carriage. Wait a few minutes, then wipe the rod clean with a soft cloth. Work the lever and spray again if necessary. Depress the lever completely and check whether the latch can hook the latch release. If the latch has worn too flat, use a small metal file to sharpen its hook so that it can catch the release *(above, right)*. Reassemble the toaster, reversing the steps taken to disassemble it.

SERVICING THE SOLENOID

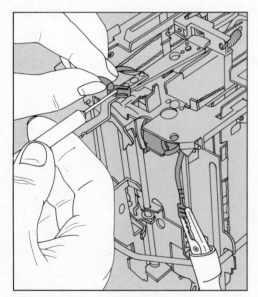

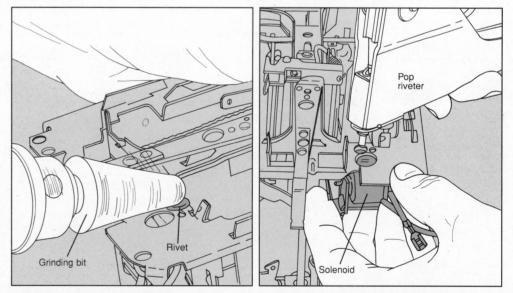

1 Testing the solenoid. Remove the chassis *(page 25)*. Pull one solenoid wire connector off the solenoid switch and pull the other off the element connecting rod. Set a multitester to RX100 *(page 113)*. Clip a tester probe to the end of one wire and touch the other probe to the end of the other wire *(above)*. The multitester should register low resistance. If the solenoid tests faulty, go to step 2 to replace it.

2 Replacing the solenoid. To remove the solenoid, wear safety goggles and use a power drill fitted with a grinding bit to grind down the mounting rivet until it is flush with the chassis *(above, left)*. Do not grind the chassis. Pry off any remaining bits of rivet with a screwdriver. Buy an exact replacement solenoid from an authorized service center and position it in the chassis so that the hole in its bracket fits right over the old rivet hole in the chassis. Wearing safety goggles, use a pop riveter fitted with a rivet the size of the bracket hole to secure the solenoid bracket to the chassis. Position the rivet head through the holes, press the riveter firmly against the chassis and squeeze its jaws closed to seat the rivet. Connect one solenoid wire to the solenoid switch terminal and the other to the element connecting-rod terminal. Reassemble the toaster, reversing the steps taken to disassemble it, then cold check for leaking voltage *(page 114)*.

TOASTER OVENS

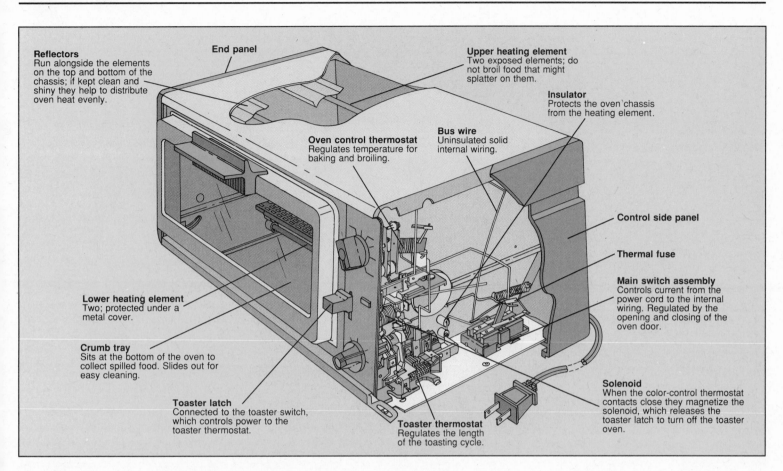

Reflectors
Run alongside the elements on the top and bottom of the chassis; if kept clean and shiny they help to distribute oven heat evenly.

End panel

Upper heating element
Two exposed elements; do not broil food that might splatter on them.

Insulator
Protects the oven chassis from the heating element.

Oven control thermostat
Regulates temperature for baking and broiling.

Bus wire
Uninsulated solid internal wiring.

Control side panel

Thermal fuse

Main switch assembly
Controls current from the power cord to the internal wiring. Regulated by the opening and closing of the oven door.

Lower heating element
Two; protected under a metal cover.

Crumb tray
Sits at the bottom of the oven to collect spilled food. Slides out for easy cleaning.

Toaster latch
Connected to the toaster switch, which controls power to the toaster thermostat.

Toaster thermostat
Regulates the length of the toasting cycle.

Solenoid
When the color-control thermostat contacts close they magnetize the solenoid, which releases the toaster latch to turn off the toaster oven.

ACCESSING THE INTERNAL COMPONENTS

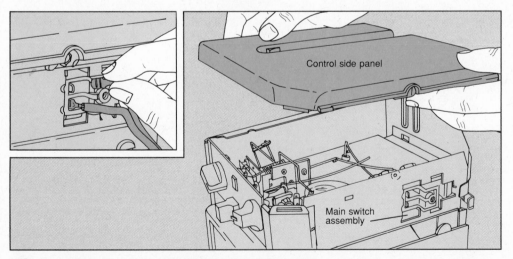

Control side panel

Main switch assembly

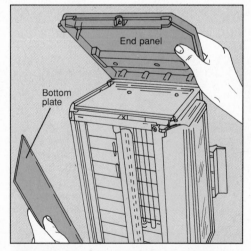

End panel

Bottom plate

1 **Removing the power cord and control side panel.** Turn off and unplug the toaster oven, and let it cool. Take the rack and broiler tray out of the oven. Trace the power cord to the main switch assembly on the bottom of the toaster oven. Turn the oven on end and remove the main switch assembly cover. Pull the power-cord wire connectors off the main switch terminals *(inset)*. You can now test the power cord *(page 116)*. To access the main switch, fuse, solenoid and heating-element terminal pins, locate and remove the screws from the control side panel. Slide the panel down to free it from the top edge tabs of the toaster oven, then lift it off the side of the oven *(above)*. To access the terminal pins on the other ends of the heating elements, go to step 2.

2 **Removing the end panel and bottom plate.** Stand the oven on its other end. Remove all screws that hold the bottom of the end panel to the bottom of the toaster oven. The model shown here has only one. Slide the panel down to free it from the top edge tabs of the toaster oven. Lift off the end panel and bottom plate *(above)*. You can now test the heating elements.

SERVICING THE MAIN SWITCH

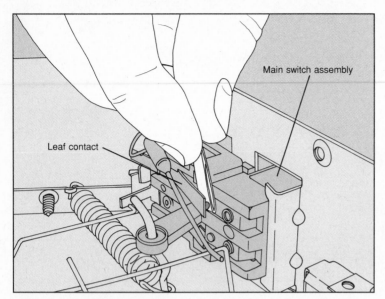

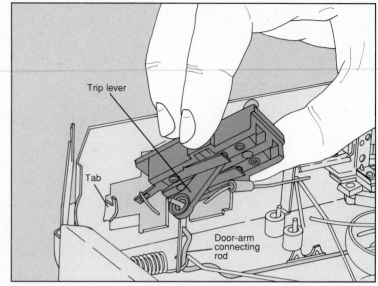

1 **Cleaning the main switch.** To access the main switch, remove the power cord and the control side panel *(page 27)*. Open and close the toaster oven door and watch whether the switch leaf contacts open and close. If the contacts are broken, bent or fused, go to step 2 to replace the main switch. If the contacts look pitted or burned, lightly rub them until they are smooth and shiny with a piece of 600-grit emery paper folded in half *(above)*. After sanding, clean the contacts by drawing plain paper between them. Reassemble the toaster oven, reversing the steps taken to disassemble it.

2 **Removing the main switch.** Use diagonal-cutting pliers to cut the bus wires as close as possible to the switch terminals. To remove the switch assembly from the toaster oven shown here, use pliers to bend back the metal tab that holds the assembly to the oven chassis, then slide the switch off its support bracket. On some models, you can simply unscrew the assembly from the chassis. Lift the switch assembly and slide its trip lever off the door-arm connecting rod *(above)*. Buy an identical switch from an authorized service center. Go to step 3 to install the new switch.

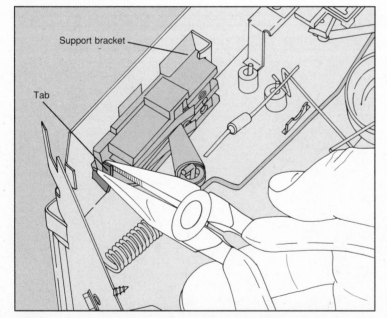

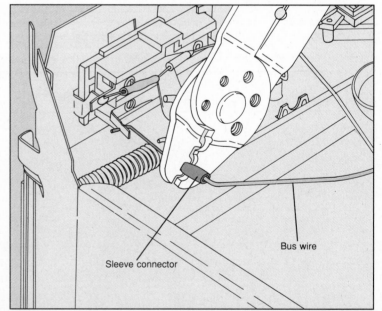

3 **Installing the main switch.** Replacement switches come with their own bus wire extensions. Slide the trip lever of the new switch assembly onto the door-arm connecting rod. Fit the assembly into its mounting hole in the toaster oven chassis and slide one end onto the support bracket. Secure the other end by bending the metal tab against it with long-nose pliers *(above)*.

4 **Connecting the main switch.** Use diagonal-cutting pliers to trim the bus wires you cut in step 2, so that they just overlap those of the new switch assembly. Install wire-to-wire uninsulated sleeve connectors *(page 119)* on the bus wires and use a multipurpose tool to crimp their ends *(above)*. Before reassembling the toaster oven, check to make sure that no sleeve connectors or bus wires are bent or touching anywhere but at their proper connection points. Reassemble the toaster oven, reversing the steps taken to disassemble it, then cold check for leaking voltage *(page 114)*.

TESTING AND REPLACING THE SOLENOID

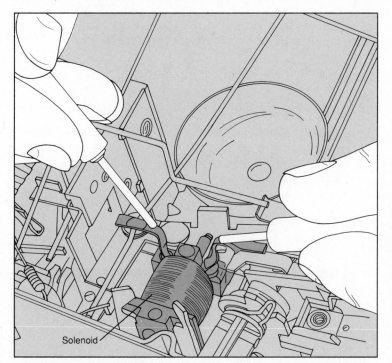

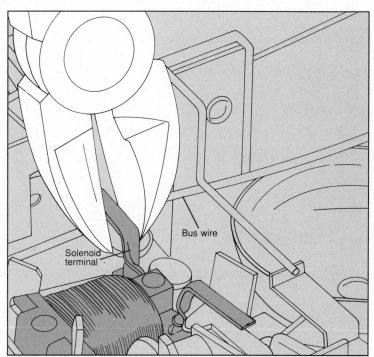

1 **Testing the solenoid.** To access the solenoid, remove the control side panel *(page 27)*. Inspect the solenoid. If the coil looks burned or if the shaft through the center cannot be moved in and out easily, go to step 2 to replace the solenoid. To test for less visible damage, set a multitester to RX100 *(page 113)*. Touch a tester probe to each solenoid terminal *(above)*. The multitester should register low resistance. If the solenoid tests faulty, replace it *(step 2)*.

2 **Replacing a faulty solenoid.** Use diagonal-cutting pliers to cut each solenoid terminal *(above)*. Use pliers to bend back the metal tabs that hold the solenoid and slide the solenoid out of its bracket on the toaster oven chassis. Purchase an exact replacement solenoid from the manufacturer or an authorized service center. A new solenoid comes with its own bus wire extension. Before you install it, cut the oven bus wire to remove the remaining bit of terminal.

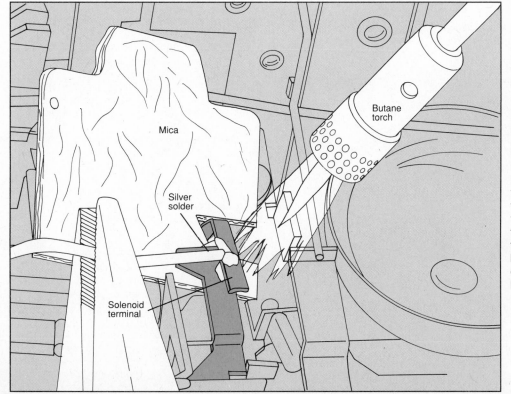

3 **Connecting the replacement solenoid.** Prepare to silver solder *(page 121)*. Slide the new solenoid into the mounting bracket on the oven chassis and bend the bracket tabs against the solenoid with pliers. Trim the bus wires of the new solenoid and the bus wires in the toaster oven so their ends just overlap. Clean the wires with emery paper to ensure a good solder. Protect the solenoid and adjacent components with a piece of mica or ceramic tile. Wearing safety glasses, use a miniature butane torch and a length of silver solder dipped in flux to solder *(page 121)* the solenoid terminal to the toaster thermostat terminal *(left)* and its bus wires to the oven bus wires. Wipe off dried flux with a foam swab and hot water. Check the solidity of the connections by squeezing them with a pair of pliers. If a connection breaks, resolder it. Reassemble the toaster oven, reversing the steps taken to disassemble it, then cold check for leaking voltage *(page 114)*.

SERVICING THE THERMAL FUSE

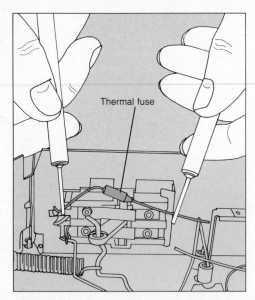

Thermal fuse

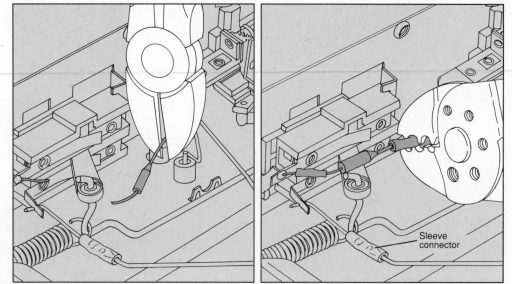

Sleeve connector

1 **Testing the thermal fuse.** Remove the control side panel *(page 27)*. On the model shown here, the fuse is located on a bus wire connected to the main switch. To test the fuse, use a continuity tester, or set a multitester to test continuity *(page 113)*. Touch a tester probe to the bus wire at each end of the fuse *(above)*. The tester should show continuity. If the fuse tests OK, take the oven for service; if faulty, replace it *(step 2)*.

2 **Replacing the thermal fuse.** Use diagonal-cutting pliers to cut the bus wires as close to the fuse as possible *(above, left)*, being careful not to twist the wires as you cut. Remove the fuse, noting the position of its pointed end for replacement, and purchase an identical fuse from an authorized service center; it will come with its own bus wires. Trim the oven bus wires so that they just overlap the bus wires of the new fuse. Install wire-to-wire uninsulated sleeve connectors to join the fuse bus wires to the oven bus wires *(page 119)*, making sure the wires overlap inside the connector. Crimp each sleeve connector with a multipurpose tool *(above, right)*. Make sure that no connectors or bus wires touch other components. Reassemble the toaster oven, reversing the steps taken to disassemble it, then cold check for leaking voltage *(page 114)*. If a replacement thermal fuse blows, suspect a faulty control thermostat and take the toaster oven for professional service.

SERVICING THE HEATING ELEMENTS

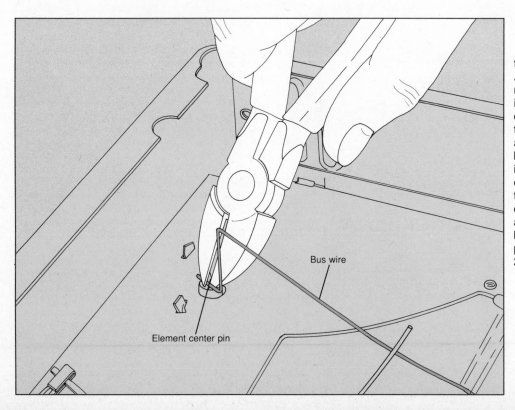

Bus wire

Element center pin

1 **Disconnecting a heating element.** Turn off and unplug the toaster oven, and let it cool. Remove the control side panel, the end panel and the bottom plate *(page 27)*. Inspect the heating elements; if an element is twisted, or burned in spots, replace it *(step 3)*. Check through both open ends of the toaster oven to inspect the connections between the center pins of the elements and their bus wires. If any connections are broken, resolder them *(step 5)*. If no damage is evident, determine which affected heating element requires replacement by removing them both from circuit for testing: Stand the oven on end with the control side facing up and use diagonal-cutting pliers to cut one bus wire of one affected element as close as possible to its center pin *(above)*. Go to step 2 to test the elements.

SERVICING THE HEATING ELEMENTS (continued)

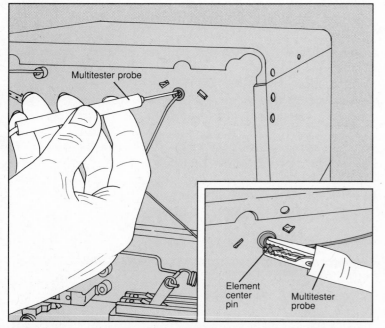

Multitester probe

Element center pin

Multitester probe

2 **Testing a heating element.** Set a multitester to test resistance *(page 113)*. Clip one tester probe to the center pin at one end of an affected element *(inset)* and touch the second probe to the other center pin of the element, at the opposite end of the toaster oven *(above)*. The multitester should show low resistance. Repeat the test for the other affected element. If both affected elements test OK, take the toaster oven for professional service. If either element or both elements test faulty, go to step 3.

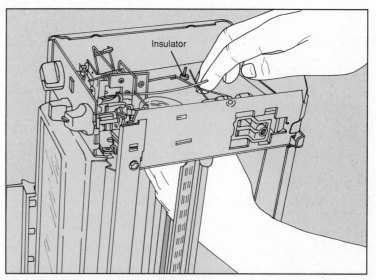

Insulator

3 **Removing a faulty heating element.** Use diagonal-cutting pliers to cut the bus wires connected to the ends of a faulty heating element, as close as possible to the element's center pins. Bend back the tabs that hold the upper element reflectors to the toaster oven frame with a pair of pliers. With one hand, reach through the oven bottom and push the side wall of the toaster oven up over the ends of the tabs *(above)*. With the other hand, push the element insulator down through the hole in the side wall. Then pull the element through the oven bottom toward you, working it free of the other side wall. If the element insulators are undamaged, remove them from the ends of the element and save them for installation on the replacement element.

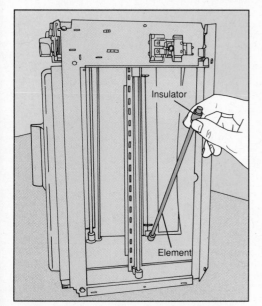

Insulator

Element

4 **Replacing a heating element.** Buy exact replacement elements, and insulators if necessary, from an authorized service center. Before installing a new element, clean its pins with a piece of emery paper and slip on the insulators. Install the new element *(above)* by reversing the steps taken to remove the faulty element, taking care not to bend it.

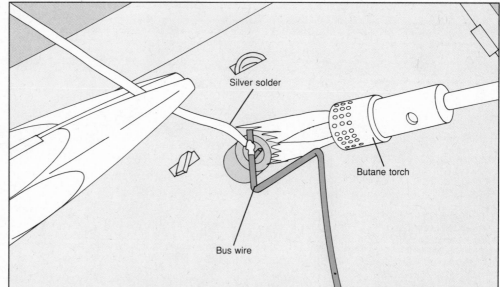

Silver solder

Butane torch

Bus wire

5 **Connecting a replacement heating element.** Prepare to silver solder *(page 121)* the oven bus wires to the heating-element center pins. Protect adjacent components with a piece of mica or ceramic tile. Clean the element center pin and bus wire tip with emery paper and push them together to make a good connection. To solder with silver solder *(page 121)*, use a miniature butane torch and a length of silver solder dipped in flux. Use the torch to heat the connection, not the silver solder. Hold the torch and solder in position until until the silver solder melts smoothly over the connection *(above)*. Turn off the torch. Repeat to connect the other end of the element. Check that all wire connections are securely soldered before reassembling the toaster oven, reversing the steps taken to disassemble it. Cold check for leaking voltage *(page 114)*.

STEAM IRONS

The steam iron produces water vapor that relaxes cloth fibers, enabling the soleplate to smooth out wrinkles. When the steam button is popped up, a valve in the reservoir opens to release droplets of water into the steam chamber. The heating element, controlled by the thermostat, is encased in a waterproof housing in the soleplate. It turns the water to steam that escapes through the soleplate ports. To take out deep wrinkles, pushing the spray button makes the pump pull water from the reservoir and expel it through the nozzle.

The irons pictured below illustrate some differences between irons made prior to the late 1970s and more recent models. Although all steam irons work on the same principles, newer models are lighter and have fewer replaceable parts. Many also have a circuit board that provides automatic shut-off. Your iron's components may differ slightly; use the irons in this chapter as a general guide to repair.

Electronic components rarely break. More likely problems are a frayed power cord or a blockage in the water system due to inadequate cleaning. If the iron is self-cleaning, use this function once a month. If not, flush the iron *(page 34)* regularly to keep it free of mineral deposits that clog spray and steam apertures. To repair your iron, consult the Troubleshooting Guide at right, then contact an authorized service center or the manufacturer to make sure parts are available and that a repair is cost-efficient. In the case of an older iron, parts may sometimes be salvaged from used irons of the same make and model; check with a small-appliance repair shop. After repair, cold-check for leaking voltage *(page 114)*.

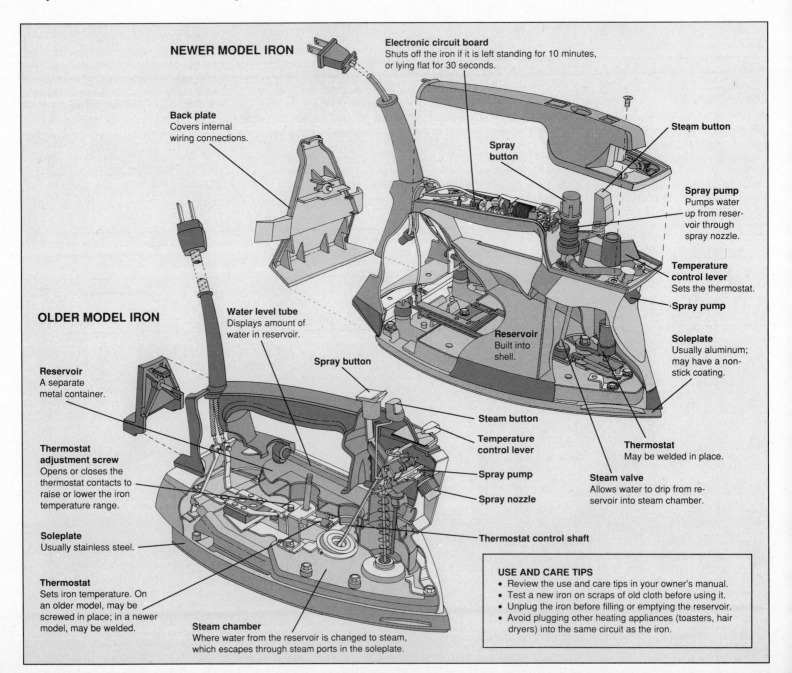

NEWER MODEL IRON

Electronic circuit board
Shuts off the iron if it is left standing for 10 minutes, or lying flat for 30 seconds.

Back plate
Covers internal wiring connections.

Spray button

Steam button

Spray pump
Pumps water up from reservoir through spray nozzle.

Temperature control lever
Sets the thermostat.

Spray pump

Soleplate
Usually aluminum; may have a non-stick coating.

Reservoir
Built into shell.

Thermostat
May be welded in place.

OLDER MODEL IRON

Water level tube
Displays amount of water in reservoir.

Spray button

Reservoir
A separate metal container.

Thermostat adjustment screw
Opens or closes the thermostat contacts to raise or lower the iron temperature range.

Soleplate
Usually stainless steel.

Thermostat
Sets iron temperature. On an older model, may be screwed in place; in a newer model, may be welded.

Steam button

Temperature control lever

Spray pump

Spray nozzle

Thermostat control shaft

Steam valve
Allows water to drip from reservoir into steam chamber.

Steam chamber
Where water from the reservoir is changed to steam, which escapes through steam ports in the soleplate.

USE AND CARE TIPS
• Review the use and care tips in your owner's manual.
• Test a new iron on scraps of old cloth before using it.
• Unplug the iron before filling or emptying the reservoir.
• Avoid plugging other heating appliances (toasters, hair dryers) into the same circuit as the iron.

TROUBLESHOOTING GUIDE

SYMPTOM	POSSIBLE CAUSE	PROCEDURE
Iron does not heat at all	Iron unplugged or turned off	Plug in and turn on iron
	No power to outlet or outlet faulty	Reset breaker or replace fuse *(p. 112)* □○; have outlet serviced
	Power cord faulty	Test and replace power cord *(p. 116)* ▣○▲
	Thermostat and/or fuse faulty	Test heating element and thermostat *(p. 35)* ▣○▲; service thermostat *(older model, p. 37* ▣◖*; newer model, p. 40* ▣○*)*
	Heating element faulty	Test heating element and thermostat *(p. 35)* ▣○▲; replace iron
	Circuit board faulty	Replace circuit board *(p. 39)* ▣○
Iron does not turn off, or becomes too hot	Thermostat calibrated incorrectly	Adjust thermostat *(p. 34)* ▣○▲
	Thermostat contacts stuck shut	Replace thermostat *(older model, p. 37; newer model, p. 40)* ▣○
Iron does not become hot enough	Power cord loose or faulty	Test and replace power cord *(p. 116)* ▣○▲
	Thermostat calibrated incorrectly	Adjust thermostat *(p. 34)* ▣○▲
	Circuit board faulty	Replace circuit board *(p. 39)* ▣○
Little or no steam	Steam ports or steam chamber aperture clogged	Clean iron *(p. 33)* □○; service pump, reservoir and steam valve assembly *(older model, p. 36; newer model, p. 39)* ▣◖
Little or no spray	Spray nozzle clogged	Service spray nozzle *(older model, p. 35; newer model, p. 39)* □○
	Pump clogged or leaking	Service pump *(older model, p. 36; newer model, p. 39)* ▣◖
Iron leaves spots or stains on fabric	Water too hard	Use distilled water or add a demineralizing product
	Soleplate, steam ports or chamber dirty	Clean iron *(p. 33)* □○
Iron gives electrical shock	Wire connections loose or broken	Remove back plate and repair wire connection *(p. 118)* ▣○
Iron leaks	Reservoir overfilled	Check owner's manual for correct filling instructions
	Spray nozzle assembly leaking	Service spray nozzle *(older model, p. 35; newer model, p. 39)* □○
	Pump gasket leaking, reservoir cracked, steam valve port clogged	Service pump, reservoir and steam valve assembly *(older model, p. 36; newer model, p. 39)* ▣◖
Water sputters from steam ports	Steam function used before iron warms up	Let iron heat for three to five minutes before using steam function
	Power cord loose or faulty	Test and replace power cord *(p. 116)* ▣○▲
Iron glides poorly or snags	Soleplate dirty or scratched	Clean soleplate *(page 33)* □○

DEGREE OF DIFFICULTY: □ **Easy** ▣ **Moderate** ■ **Complex**
ESTIMATED TIME: ○ **Less than 1 hour** ◖ **1 to 3 hours** ● **Over 3 hours**   ▲ **Special tool required**

CLEANING THE IRON

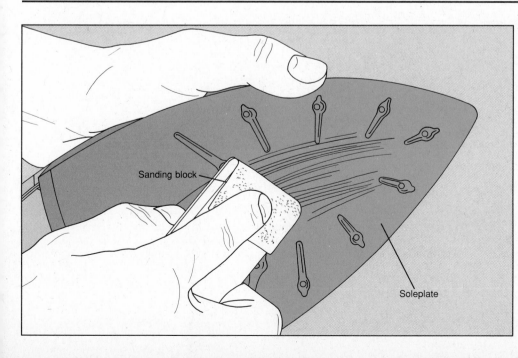

Sanding block

Soleplate

1 Cleaning the soleplate. Heat the iron, then turn it off and unplug it. Let it cool just enough to handle. Wipe a non-stick soleplate with a soft, damp cloth or sponge. To remove residue and scratches from aluminum and stainless steel soleplates, use a piece of 600-grit waterproof emery paper wrapped around a small block of wood. Sprinkle a bit of water on the soleplate and sand it with a steady back-and-forth motion, heel to tip, to produce an even, dull sheen *(left)*. Wipe the sanded areas clean with a damp cloth. To finish off, buff the soleplate with 4/0 steel wool until it has a smooth satin finish. Wipe it again with a damp cloth to remove steel wool particles.

CLEANING THE IRON (continued)

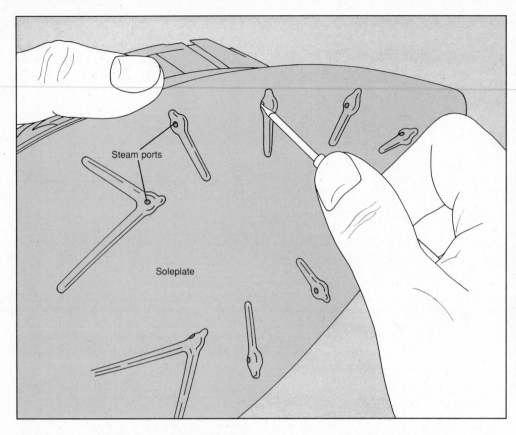

Steam ports

Soleplate

2 **Unclogging the steam ports and flushing the iron.** Heat the iron, then turn it off and unplug it. Let it cool just enough to handle. With the tip of a small screwdriver, scrape out mineral deposits encrusting the edges of the steam ports *(left)*. Take care not to scratch the soleplate, and hold the soleplate at an angle to prevent scrapings from falling into the steam chamber. Soften stubborn deposits by moistening them with a cotton swab dipped in vinegar and letting them soak for 15 minutes. After cleaning the steam ports, use the iron's self-cleaning function according to the directions in the owner's manual. If the iron is not self-cleaning, flush it clean: Mix a half-and-half solution of vinegar and water and fill the reservoir. Stand the iron upright in a sink. Set the steam button to OFF, in most cases by depressing it and locking it down. Plug in the iron, set the temperature at its highest setting and let the iron heat for five minutes. Turn off the iron and unplug it. Lay the iron flat in the sink, release the steam button and let the steam and water pour out until the iron is empty, usually 5 to 10 minutes. Repeat twice with clear water to flush the iron completely.

ADJUSTING THE THERMOSTAT

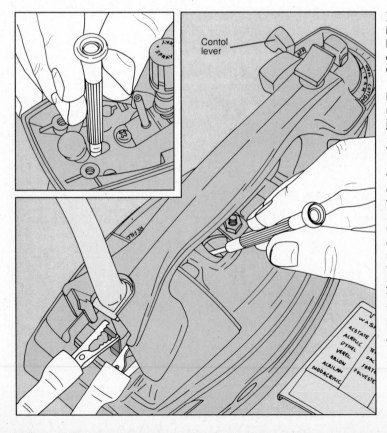

Contol lever

Performing an ON/OFF calibration test. Turn off and unplug the iron. Let it cool to room temperature. Unscrew the back plate, and label and disconnect the wires from the main terminal posts. To check whether the thermostat is correctly calibrated in both the OFF and ON positions, set a multitester to test resistance *(page 113)* and clip a multitester probe to each terminal. Set the control lever to OFF; the multitester should indicate infinite resistance. If not, adjust the thermostat contacts. To access the thermostat adjustment screw on many older irons, pry off the saddle plate *(page 36, step 3)*. On newer irons you may need to pull off the temperature button on the side of the shell or remove the handle plate *(page 39, step 1)* to reach the screw *(inset)*. Turn the screw clockwise *(left)*, one-quarter turn at a time, until the thermostat contacts open and the multitester measures infinite resistance. If the contacts are stuck shut, replace the thermostat *(older model, page 37; newer model, page 40)*. Once you have infinite resistance in the OFF position, check the thermostat in the ON position: Turn the control lever to the lowest possible ON setting. The contacts should close and the multitester should indicate about 12 ohms. If the multitester continues to show infinite resistance, turn the adjustment screw counterclockwise, one-quarter turn at a time, until the multitester shows about 12 ohms. Double-check to be sure that there is still infinite resistance in the OFF position. Test the iron on scraps of fabric before ironing clothes. If you had to turn the adjustment screw more than one full turn for either test, take the iron for professional service. If an iron with an automatic shutoff still doesn't heat up enough after the ON/OFF calibration test, suspect a faulty circuit board and replace it *(page 39)*.

TESTING THE HEATING ELEMENT AND THERMOSTAT

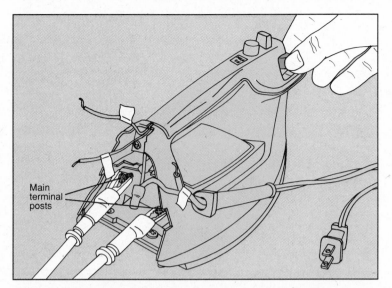

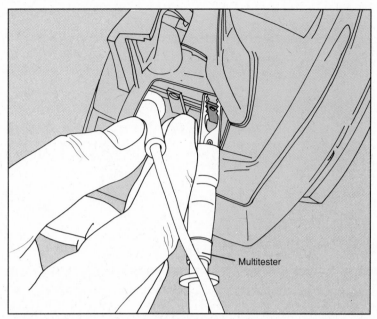

1 **Testing the heating element and thermostat together.** Turn off and unplug the iron, and let it cool. Unscrew the back plate. If the iron does not have an automatic shutoff, go to step 2 to diagnose the electrical problem. On a model with automatic shutoff, locate the main terminal posts for the heating element and thermostat; one is connected to a power cord wire and the second to a circuit board wire. Label and disconnect the wires. Set a multitester to test resistance *(page 113)* and clip a probe to each terminal post *(above)*. Turn on the iron. The multitester should register partial resistance. If not, go to step 2. If the iron tests OK, replace the circuit board *(page 39)*.

2 **Testing the heating element alone.** Locate the two heating element terminals on the far left and far right edges of the opening. Set a multitester to test resistance *(page 113)*. Clip one probe to one terminal and touch the second probe to the other terminal *(above)*. The multitester should register partial resistance. If the heating element tests OK, suspect a faulty thermostat *(older model, page 37; newer model, page 40)*. If the heating element tests faulty, replace the iron.

SERVICING THE SPRAY NOZZLE AND FRONT PLATE ASSEMBLY (Older model)

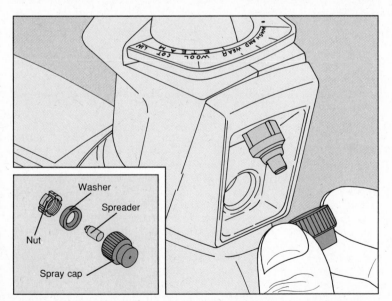

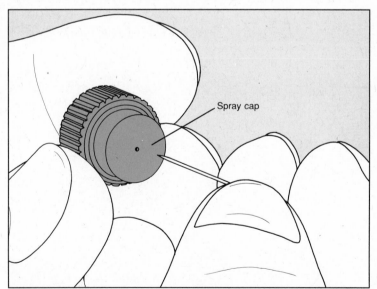

1 **Removing the spray nozzle assembly.** Turn off and unplug the iron. Let it cool to room temperature before removing the spray nozzle assembly. Unscrew and remove the spray cap *(above)*. Use long-nose pliers to loosen the cap if it is stuck. Pull off the spreader underneath, and the washer behind it. Then unscrew and remove the nut that holds the front plate in position. Note the position of the spray nozzle assembly parts for correct reassembly *(inset)*.

2 **Cleaning the spray nozzle assembly.** To service the spray nozzle assembly, inspect the center hole of the spray cap. If it is clogged, clear it with the tip of a fine needle *(above)*. Do not force the needle or you may enlarge the hole, damaging the spray action. Replace a dry or cracked washer or a damaged spray cap, spreader or nut with exact replacement parts. Reassemble the spray nozzle and flush the iron thoroughly *(page 34)*.

SERVICING THE PUMP, RESERVOIR AND STEAM VALVE ASSEMBLY (Older model)

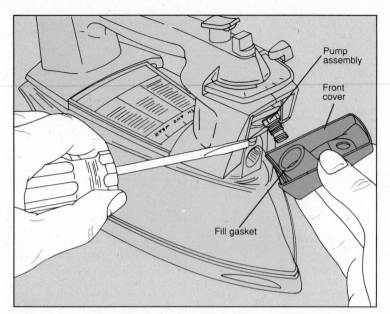

Pump assembly

Front cover

Fill gasket

1 **Removing the front cover and replacing the fill gasket.** Turn off and unplug the iron, and let it cool. If you suspect a damaged pump, leaking reservoir or clogged steam-chamber aperture, disassemble the iron *(steps 1-4)*. First, remove the spray nozzle assembly *(page 35)*. Then remove the front cover by slipping a screwdriver blade into one of the notches on the side of the front cover and prying it loose *(above)*. Replace a dried or cracked fill gasket with an exact replacement from a small-appliance repair shop or the manufacturer. Reinstall the front cover and spray nozzle assembly *(page 35)*, or go to step 2.

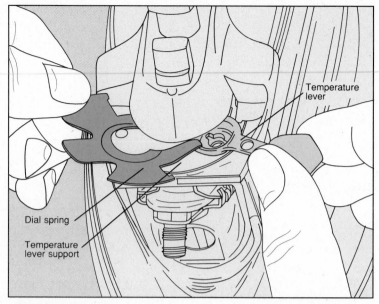

Temperature lever

Dial spring

Temperature lever support

2 **Removing the temperature selector assembly.** Slip a screwdriver blade under the front edge of the handle in the center of the dial plate. Press down on the dial plate to release it. Slide the dial plate forward and out from under the edge of the handle. Pull the dial spring, the temperature lever and the temperature lever support out from beneath the handle *(above)*. Note the position of the temperature-selector assembly parts for correct reassembly *(page 38, step 4)*.

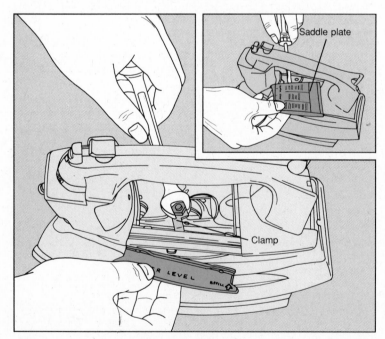

Saddle plate

Clamp

3 **Removing the saddle plate and water level plates.** Insert the screwdriver blade under the edge of the saddle plate and pry it off *(inset)*; the thermostat adjustment screw is now visible through the hole in the shell. To remove the water level plate, use a small open-end wrench to loosen the nut that holds the clamp over the edge of the water level plate. Slide the plate out from under the glass water-level tube *(above)*. Unscrew and remove the nut and clamp.

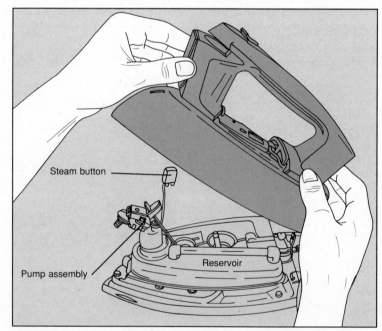

Steam button

Pump assembly

Reservoir

4 **Removing the handle and shell.** Unscrew the back plate and label and disconnect the power cord wires. Lift off the shell and handle *(above)* to reveal the pump assembly and reservoir. The steam button may remain attached to the reservoir or it may be pulled loose. Set the steam button and rod aside. Note the location of parts for correct reassembly. To clean or replace the pump, go to step 5. To replace the reservoir, clean the steam chamber aperture or service the thermostat, go to step 6.

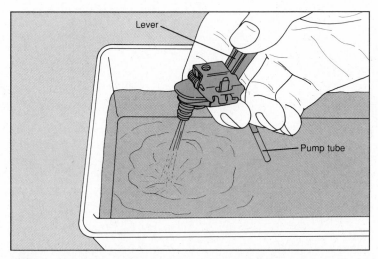

Lever

Pump tube

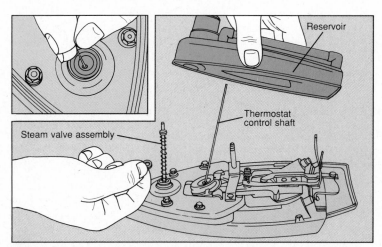

Reservoir

Thermostat control shaft

Steam valve assembly

5 **Servicing the pump.** Pull the pump out of the reservoir. If you suspect the pump leaks, fill a container with water, insert the pump tube into the water and depress the lever a few times. If the pump is clogged and does not squirt water forcefully, clean it. Soak the pump in vinegar for 15 to 20 minutes to soften mineral buildup, then test it again. If the pump leaks or remains clogged after cleaning, purchase an exact replacement pump from a small-appliance repair shop or from the manufacturer. Install the pump in its hole in the reservoir and reassemble the iron *(page 38)*, or go to step 6.

6 **Replacing the reservoir and cleaning the steam chamber aperture.** Pull the small gasket off the top of the steam valve assembly. Unscrew the front clamp that holds the reservoir to the soleplate, and lift it off. Store the small gasket on the tip of the steam valve assembly *(above)*. Replace a reservoir that is cracked or corroded with a duplicate from a small-appliance repair shop or the manufacturer. Pull the steam valve assembly off the soleplate. Inspect the tiny aperture into the steam chamber; if clogged with mineral deposits, open it with a straight pin *(inset)*. Examine the thermostat *(step below)*. After completing repair, reassemble the iron *(page 38)*.

SERVICING THE THERMOSTAT (Older model)

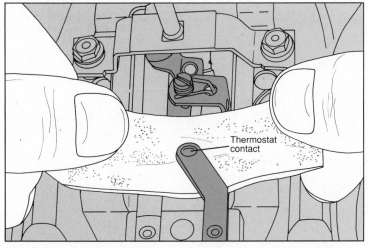

Thermostat contact

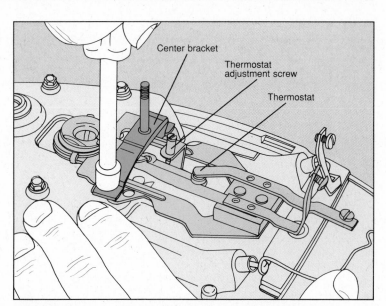

Center bracket

Thermostat adjustment screw

Thermostat

1 **Inspecting and cleaning the thermostat.** Turn off the iron, unplug it and allow it to cool. Unscrew the back plate, and label and disconnect the power cord wires. Remove the spray nozzle assembly *(page 35)* and the pump and reservoir *(page 36)*. Turn the thermostat control shaft back and forth and watch whether the contacts open and close. If they are stuck shut, replace the thermostat *(step 2)*. If the contacts look pitted or black, use a piece of 600-grit emery paper to smooth and polish them, then clean them with plain paper. Place the paper between the contacts, turn the control shaft to close them and pull the paper back and forth very gently *(above)*. After cleaning, check that the thermostat contacts open and close properly when the control shaft is turned. Reassemble the iron *(page 38)*. If the thermostat cannot be cleaned, replace the assembly *(step 2)*.

2 **Replacing the thermostat.** Buy an identical replacement assembly from a small-appliance repair shop or the manufacturer. Pull off the thermostat control shaft, noting its exact position. Disconnect the heating element from the thermostat. Unscrew the center bracket and thermostat from the soleplate *(above)*. Remove any washers and the center bracket. Unscrew the thermostat adjustment screw as far as you can without removing it. Slide the thermostat assembly toward the rear until you can lift it up over the adjustment screw bracket and off the soleplate. Position the new thermostat assembly on the soleplate and reassemble the iron *(page 38)*. Adjust the thermostat *(page 34)*.

REASSEMBLING THE IRON (Older model)

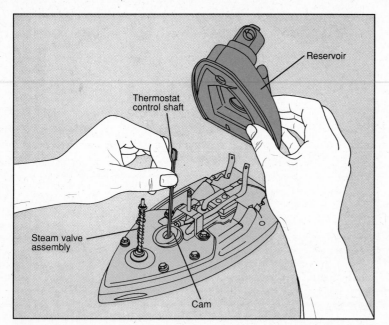

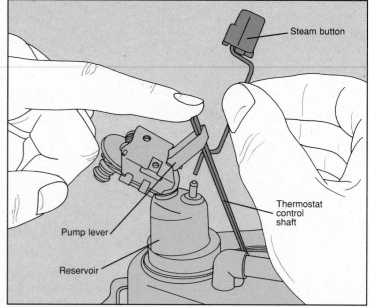

1 **Reinstalling the reservoir.** Fit the steam valve assembly into the steam valve seat, pressing the spring firmly in place so it stands upright. Take the small gasket off its tip. Reinstall the thermostat control shaft *(above)*, aligning the ridge on its side with the slot in the thermostat cam. Position the reservoir, feeding the steam valve assembly and thermostat control shaft through their holes in the front of the reservoir. Place the front clamp over the front edge of the reservoir and screw it to the soleplate below.

2 **Reinstalling the pump and steam knob.** Slip the small gasket onto the tip of the steam valve assembly. Reposition the pump in the reservoir, inserting the thermostat control shaft through its lever. Insert the steam button rod through the opening in the steam valve assembly *(above)*. Position the steam button so that you can read the lettering on the button from the back of the iron, and so that the bend in the rod fits behind the pump lever.

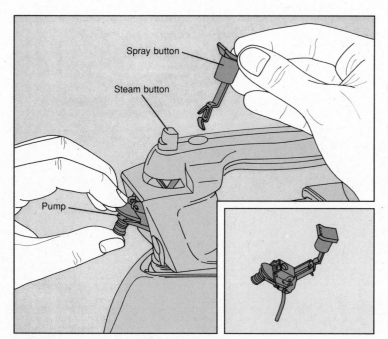

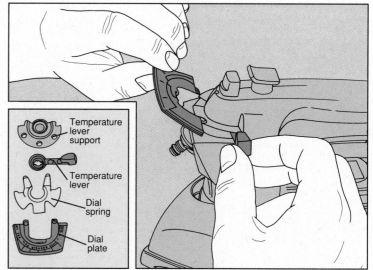

3 **Reinstalling the shell and spray button.** Working from the front of the iron, lower the shell over the iron with one hand. With the fingers of your other hand, guide the steam button up through its hole in the handle. If necessary, insert a screwdriver into the spray button hole to help push the steam button into place. Once the shell is in position, insert the spray button shaft into the handle *(above)*. Maneuver the button and the pump until the shaft clamps securely onto the pump lever inside *(inset)*.

4 **Reinstalling the temperature selector assembly.** Check the correct order of the temperature-selector assembly parts *(inset)*. Reinstall them one by one, beginning with the temperature lever support and finishing with the dial plate *(above)*. Fit the lever and control over the top of the temperature control shaft. Install the fill gasket and front cover. Check the proper order of the spray-nozzle assembly parts *(page 35, step 1)* and reinstall them. Resecure the water level plate with the clamp and nut. Before replacing the saddle plate, perform an ON/OFF calibration test *(page 34)*. Snap on the saddle plate, reconnect the power cord and screw on the back plate. Cold check for leaking voltage *(page 114)*. Clean and flush the iron thoroughly *(page 34)*.

SERVICING THE SPRAY NOZZLE AND PUMP ASSEMBLY (Newer model)

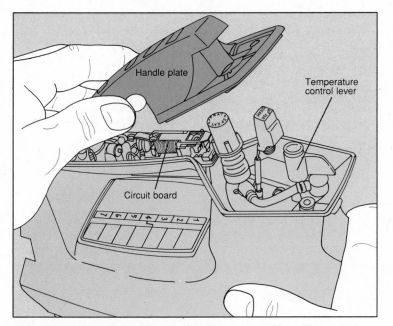

Handle plate

Temperature control lever

Circuit board

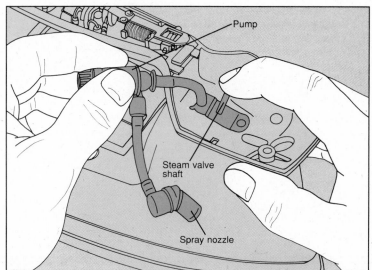

Pump

Steam valve shaft

Spray nozzle

1 **Removing the handle plate.** Turn off and unplug the iron, and let it cool. Turn the temperature lever to OFF. Push the dial plate back slightly with your fingers, then slip the blade of a screwdriver under its edge and pry it off. Unscrew and remove the two screws normally hidden under the dial plate. Turn the temperature control lever to the center position, aligning it with the slot in the center of the handle, and lift off the handle plate *(above)*.

2 **Checking and replacing the spray nozzle and pump.** Pull the spray nozzle free and use the tip of a fine needle to clear its aperture of mineral deposits. (Do not force the needle or you may create a leak.) Pull out the temperature lever, the spray button, and the steam button and its rod. Remove the screw that holds the pump bracket in place. Steady the steam valve shaft with a finger while you slide the pump out from under the edge of the bracket with the other hand *(above)*. Test the pump for leaks and clean it *(page 37, step 5)*. Replace a damaged pump with an exact replacement purchased from a small-appliance repair shop or the manufacturer. Install the new pump and reassemble the iron, reversing these steps.

REPLACING THE ELECTRONIC CIRCUIT BOARD (Newer model)

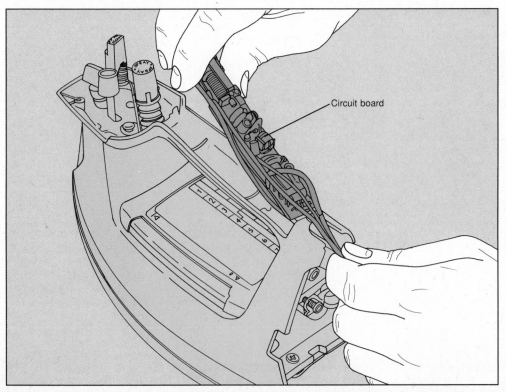

Circuit board

Taking out the circuit board. Unscrew and remove the back plate. Label and disconnect the power cord wires and circuit board wires; unscrew screw-on connectors and use a pair of diagonal-cutting pliers to cut off any crimp connectors. If necessary, draw a diagram to help you remember where the wires go. Remove the handle plate *(step above)*. Lift the circuit board out of its seat in the handle *(left)* and inspect it for cracks or evidence of water damage. Buy an exact replacement from an authorized service center or the manufacturer. Position the new circuit board in the handle. Reconnect the power cord wires and circuit board wires; if crimp connectors were used, crimp on new connectors *(page 119)*. Reinstall the back plate, the handle plate and the dial plate, reversing the steps you took to remove them.

SERVICING THE THERMOSTAT (Newer model)

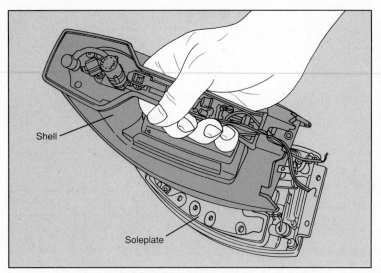

Shell

Soleplate

1 **Removing the shell.** Unscrew and remove the back plate. If necessary, draw a diagram to help you remember where the wires go, then disconnect the power cord wires and any circuit board wires. Unscrew screw-on connectors and use a pair of diagonal-cutting pliers to cut off any crimp connectors. Remove the handle plate of the iron *(page 39)*. Pull out the temperature lever and the steam button and its rod. To separate the shell from the soleplate beneath, unscrew the two screws that hold the shell to the heel of the soleplate, and loosen (but do not remove) the two recessed screws that secure the shell to the tip of the soleplate. Lift the shell off the soleplate *(left)*.

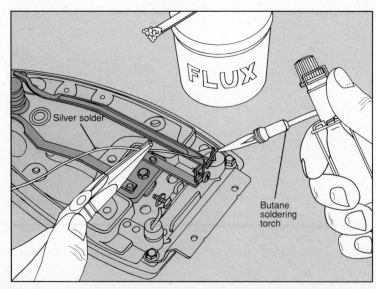

Fuse

Thermostat assembly

Heating element wire

Contacts

2 **Removing the thermostat assembly.** Inspect the thermostat contacts and clean them *(inset)* if necessary *(page 37, step 1)*. If the thermostat fuse has opened, or damage to the contacts is severe, replace the thermostat assembly. Most newer irons, such as the one shown, have a thermostat that is spot-welded to the heating element. When replaced, this type must be silver soldered; have the right tools and materials on hand *(page 121)*. Buy an exact replacement thermostat assembly from a small-appliance repair shop or the manufacturer. Lift the temperature control-arm seat off the front of the old thermostat assembly. Use diagonal-cutting pliers to sever the small piece of wire that connects the metal arm of the thermostat assembly to the heating element terminal *(left)*, cutting 1/8 inch from the terminal. Unscrew the thermostat from the soleplate. On the model shown, one screw is located at the back of the assembly on a small porcelain insulator, and the other is located in the front, on the thermostat. Lift off the old thermostat assembly.

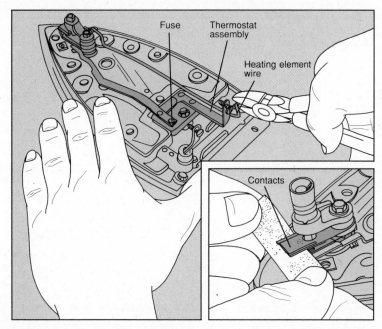

FLUX

Silver solder

Butane soldering torch

3 **Soldering the new thermostat assembly.** Lay the new thermostat assembly in position on the soleplate, and screw it in place. The metal arm of the new assembly should just touch the cut end of the wire connected to the heating element terminal. If the wire overlaps the arm of the thermostat assembly, cut it shorter with the diagonal-cutting pliers. Use a utility knife to make a vertical incision through the rubber insulator wrapped around the heating element terminal, and remove the insulator. Silver solder *(page 121)* the arm of the thermostat assembly to the heating element *(left)*. Wait five minutes for the area to cool. Slip the rubber insulator back around the heating element terminal. Reposition the temperature control-arm seat at the front of the thermostat assembly. To reassemble the iron, screw on the shell and reconnect the power cord wires and any circuit board wires *(page 118)*. Perform an ON/OFF calibration test *(page 34)* and adjust the thermostat if necessary. Then replace the temperature control arm, steam button and rod, spray nozzle and pump. Screw on the back plate and reinstall the handle plate and the dial plate. Do a cold check for leaking voltage *(page 114)*.

COFFEE MAKERS

The ritual of morning coffee rivals Mom and apple pie, and a kitchen without some sort of coffee machine hardly seems complete. The two most typical coffee makers are the automatic steam-pump drip brewer and the percolator. Automatic drip coffee makers *(page 42)* usually have one heating element controlled by a thermostat, which cycles it on and off until the unit is unplugged or it shuts off automatically. An electronic shutoff feature may be found with a digital timer that can be set to start the brewing cycle at a certain time. Percolators *(page 46)* have separate brewing and warming elements. Both elements go on the moment the unit is plugged in, but when the coffee gets hot enough, the thermostat cuts power to the brewing element. The warming element stays on as long as the unit is plugged in.

When your coffee maker malfunctions, use the Troubleshooting Guide below to diagnose the problem. On newer models, thermostat and element connections may be spot welded; be prepared to silver solder new connections by reading the instructions in Tools & Techniques *(page 121)*. If the coffee is too strong, too weak or bitter, don't immediately blame the coffee maker. Often such problems can be traced to the amount or type of coffee being used or inadequate cleaning. At least once a month, run an automatic drip coffee maker through a brewing cycle with a half-and-half solution of vinegar and water and then with cold water. Rinse a percolator's pump tube and basket with water after every use. To remove coffee stains from the basket, soak it in boiling water with two tablespoons of dishwasher powder, then rinse well.

TROUBLESHOOTING GUIDE

SYMPTOM	POSSIBLE CAUSE	PROCEDURE
AUTOMATIC DRIP COFFEE MAKERS		
Coffee maker doesn't work at all	No power to outlet or outlet faulty	Reset breaker or replace fuse *(p. 112)* □○; have outlet serviced
	Power cord faulty	Test and replace power cord *(p. 116)* ◼○▲
	Fuse or thermostat faulty	Test and replace fuse and/or thermostat *(p. 43)* ◼○
	Heating element faulty	Test and replace heating element *(p. 44)* ◼○▲
	Wire connection loose or broken	Repair wire connection *(p. 118)* ◼○
	On/off switch faulty (non-electronic)	Test and replace on/off switch *(p. 45)* ◼○▲
	Circuit board faulty (electronic)	Replace circuit board *(p. 45)* ◼○
Coffee maker sputters and brews coffee slowly	Steam pump tubes clogged	Consult owner's manual and clean tubes
Coffee maker overflows	Reservoir overfilled	Consult owner's manual and fill reservoir correctly
	Filter basket clogged or overfilled	Consult owner's manual; clean and fill basket correctly
Coffee maker leaks	Reservoir or steam pump tubes cracked	Take coffee maker for professional service
Coffee maker blows circuit breaker or fuse	Wire connection loose or broken	Repair wire connection *(p. 118)* ◼○
	Power cord faulty	Test and replace power cord *(p. 116)* ◼○▲
PERCOLATORS		
Percolator doesn't work at all	No power to outlet or outlet faulty	Reset breaker or replace fuse *(p. 112)* □○; have outlet serviced
	Power cord faulty	Test and replace power cord *(p. 116)* ◼○▲
	Power-cord receptacle terminal pins faulty	Service terminal pins *(p. 46)* ◼○
	Wire connection loose or broken	Repair wire connection *(p. 118)* ◼○
	Fuse or brewing element faulty	Take percolator for professional service
Percolator heats water but doesn't perk coffee	Steam tube or basket clogged	Consult owner's manual and clean tube and basket
	Thermostat faulty	Test and replace thermostat *(p. 47)* ◼○
Perked coffee boils and reperks	Thermostat faulty	Test and replace thermostat *(p. 47)* ◼○
	Warming element faulty	Test and replace warming element *(p. 47)* ◼○▲
Perked coffee doesn't stay hot	Warming element faulty	Test and replace warming element *(p. 47)* ◼○▲
Percolator blows circuit breaker or fuse	Power cord faulty	Test and replace power cord *(p. 116)* ◼○▲
Percolator gives electrical shock	Wire connection loose or broken	Repair wire connection *(p. 118)* ◼○
Percolator leaks	Seals or gaskets worn	Take percolator for professional service

DEGREE OF DIFFICULTY: □ Easy ◼ Moderate ◼ Complex
ESTIMATED TIME: ○ Less than 1 hour ◖ 1 to 3 hours ● Over 3 hours

▲ Special tool required

AUTOMATIC DRIP COFFEE MAKERS

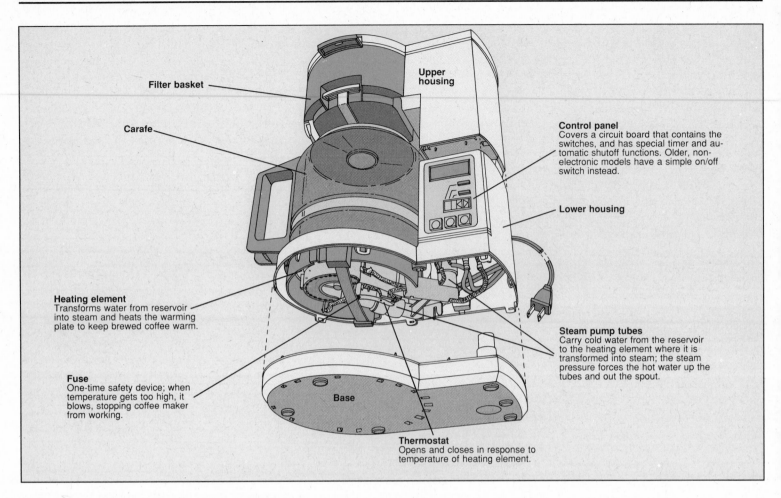

Filter basket

Upper housing

Carafe

Control panel
Covers a circuit board that contains the switches, and has special timer and automatic shutoff functions. Older, non-electronic models have a simple on/off switch instead.

Lower housing

Heating element
Transforms water from reservoir into steam and heats the warming plate to keep brewed coffee warm.

Steam pump tubes
Carry cold water from the reservoir to the heating element where it is transformed into steam; the steam pressure forces the hot water up the tubes and out the spout.

Fuse
One-time safety device; when temperature gets too high, it blows, stopping coffee maker from working.

Base

Thermostat
Opens and closes in response to temperature of heating element.

ACCESSING INTERNAL COMPONENTS

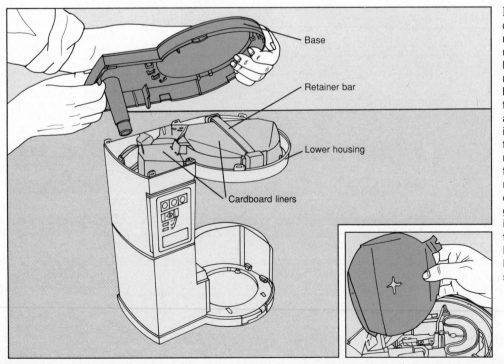

Base

Retainer bar

Lower housing

Cardboard liners

Removing the base. Turn off and unplug the coffee maker and let it cool. Remove the carafe and the filter basket and turn the unit upside down. Locate and remove the screws holding the base to the lower housing. On many models, you may then simply lift off the base. On the model shown here, gently work the base off the lower housing with the tip of a screwdriver and then lift it off *(left)*. If the internal components are covered by protective cardboard liners, remove them. On the model shown here, pull out the liner that protects the circuit board assembly, taking care not to tug on any wires. To remove the liner covering the heating element assembly, gently push down the retainer bar and rotate it clockwise to slip its ends out of the slots in the heating element housing. Remove it, then lift off the cardboard liner *(inset)*. You now have access to the heating element, thermostat, thermal fuse and circuit board.

TESTING AND REPLACING THE FUSE AND THERMOSTAT

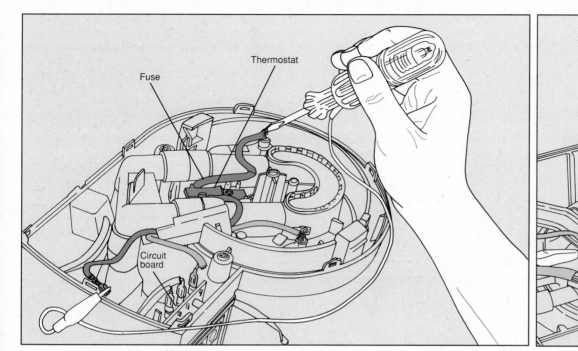

1 **Testing the fuse and thermostat.** To gain access to the thermostat and fuse, turn off and unplug the unit and remove the base *(page 42)*. Locate the small disc thermostat and the thick rubber- or plastic-covered fuse connected to one of its terminals. If you have a coffee maker in which the thermostat and fuse are spot welded together, disconnect the wire that connects the thermostat to the circuit board and test thermostat and fuse as a single assembly. Touch one continuity tester probe to the end of the free wire and touch the other to the end of the fuse wire where it is spot welded to a heating element terminal *(above, left)*. The tester should light. If the assembly tests faulty, replace it *(step 2)*. If the thermostat and fuse in your coffee maker are connected with wire terminal connectors, test them separately. To test a fuse, disconnect one of its wires and touch a continuity tester probe to each wire end. To test a thermostat, disconnect the wire from one of its terminals and touch a tester probe to each terminal *(above, right)*. In both cases, the tester should light. If either the fuse or the thermostat tests faulty, replace it *(step 2)*.

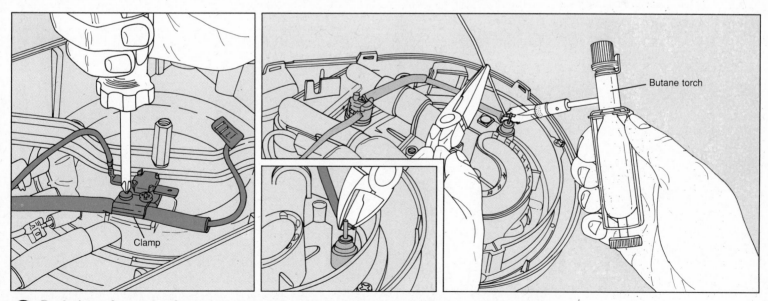

2 **Replacing a fuse and a thermostat.** To remove either a thermostat-and-fuse assembly or a separate thermostat or fuse, unscrew any clamps securing them to the element *(above, left)*. Disconnect their wires; if spot welded, use diagonal-cutting pliers to cut them loose near the terminal *(inset)*. Buy identical replacement parts from an authorized service center and install them, reversing the steps you took to remove the faulty ones. Silver solder *(page 121)* wires that were spot welded; wrap the wire around the terminal and use silver solder and a miniature butane torch *(above, right)*. Reassemble the coffee maker and cold check for leaking voltage *(page 114)*.

TESTING AND REPLACING THE HEATING ELEMENT

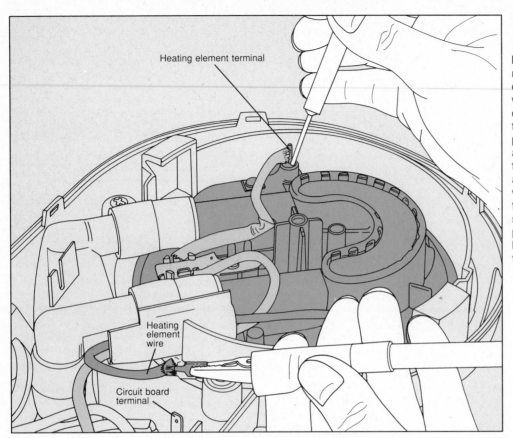

Heating element terminal

Heating element wire

Circuit board terminal

1 **Testing the heating element.** To gain access to the heating element, turn off and unplug the unit and remove the base *(page 42)*. Before testing the element, remove it from circuit: Either pull the wire off one of its terminals or, if the wire is spot welded as in the model shown here, pull it off the circuit board terminal. Set a multitester to RX1 *(page 113)*. Touch a tester probe to each heating element terminal or, as in the model shown here, touch one tester probe to the heating element terminal and the other to the end of the disconnected wire *(left)*. The multitester should show partial resistance. If the element tests OK, consult the Troubleshooting Guide for other possible causes of coffee maker malfunction before reassembling the unit. If the element tests faulty, go to step 2 to replace it.

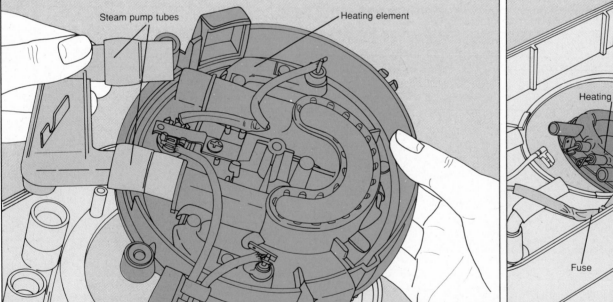

Steam pump tubes

Heating element

Thermostat

Heating element

Fuse

2 **Replacing the heating element.** Replacement heating elements for the model shown on page 42 come as an assembly with the fuse and thermostat together. To remove the heating element, fuse and thermostat together, pull the thermostat wire off the circuit board terminal, then unscrew the heating element assembly from the lower housing. Lift the assembly off the lower housing with one hand while you pull the rubber steam-pump tubes off it with the other *(above, left)*. On many other coffee makers, the heating element may be replaced separately; label and dis-

connect the wires attached to the heating element terminals, and remove the thermostat and fuse from the element by unscrewing their mounting clamp. Then remove the center post that secures the element retainer bracket and lift off the bracket. Pull the steam pump hoses off the element and remove the element from the housing *(above, right)*. Buy exact replacement parts from an authorized service center, and reinstall them, reversing the steps taken to remove them. Reassemble the coffee maker and cold check for leaking voltage *(page 114)*.

REPLACING THE CIRCUIT BOARD

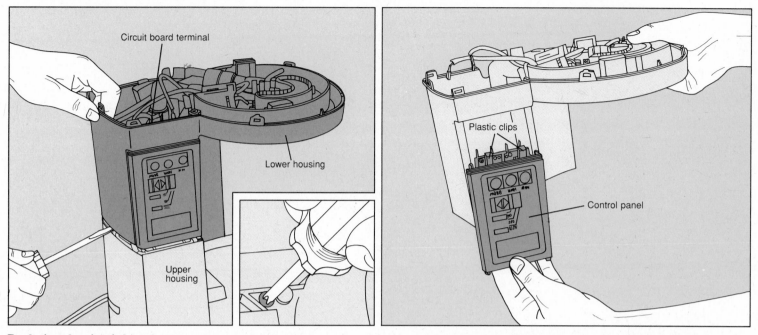

Circuit board terminal

Lower housing

Upper housing

Plastic clips

Control panel

Replacing the circuit board. If the on/off function is faulty on an electronic coffee maker, replace the circuit board. To access the circuit board, turn off and unplug the unit and remove the base *(page 42)*. Label all wires and pull them off the circuit board terminals. Locate and remove any screws securing the circuit board to the housing. On the model shown here, reach behind the circuit board and remove the screw that holds the lower housing to the upper housing *(inset)*. Pull the ends of the steam pump tubes out of the rubber sleeves that hold them to the upper housing. Then

separate the lower and upper housing by working the tip of a screwdriver along the seam between the two sections *(above, left)*. Lift off the lower housing. Reach inside the housing with a screwdriver and pry back the plastic clips that secure the circuit board to the housing. Slide the circuit board (with the control panel attached) out of the lower housing *(above, right)*. Purchase a replacement circuit board from an authorized service center and install it by reversing the steps taken to remove the faulty one. Reassemble the coffee maker and cold check for leaking voltage *(page 114)*.

SERVICING THE ON/OFF SWITCH

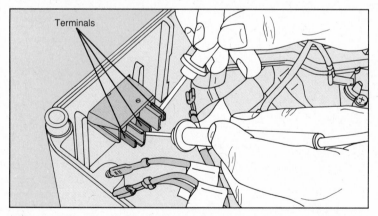

Terminals

1 Testing the on/off switch. Coffee makers without an electronic timer or automatic shutoff have a simple on/off switch. Turn off and unplug the unit, let it cool and remove its base *(page 42)*. Locate the switch, label its wires and pull them off the terminals. A typical on/off switch has a pilot light and three terminals. Set a multitester to RX1 *(page 113)*. Set the switch to OFF and touch the tester probes to each possible pairing of terminals *(above)*, in turn. The multitester should never indicate 0 ohms, although it may show low resistance between one pair of terminals. Set the switch to ON and repeat the test. The tester should register 0 ohms for only one pair of terminals. If the switch fails either test, go to step 2 to replace it.

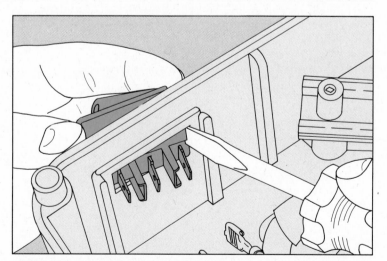

2 Replacing the on/off switch. Before removing the switch, note which side faces up so you can install the replacement the same way. To remove it, press in the retaining clips on one side with a screwdriver blade and angle it out through the housing; repeat on the other side until you can push the switch out completely *(above)*. Buy an identical switch from an authorized service center. Snap the new switch into the housing and reconnect the wires. Reassemble the coffee maker, reversing the steps taken to disassemble it, and cold check for leaking voltage *(page 114)*.

PERCOLATORS

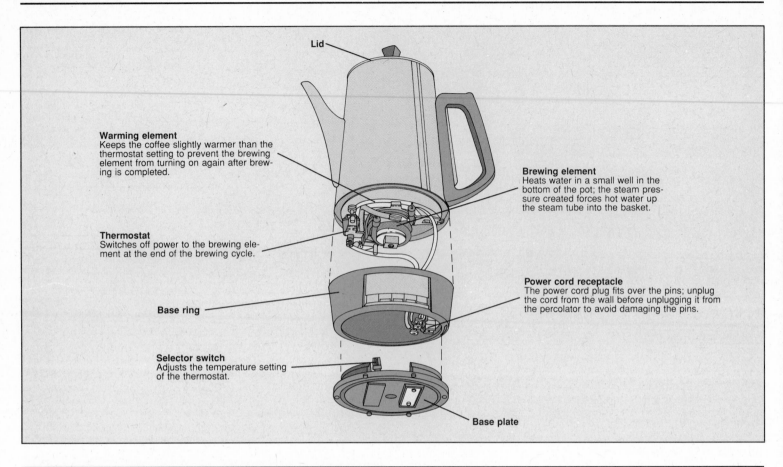

Lid

Warming element
Keeps the coffee slightly warmer than the thermostat setting to prevent the brewing element from turning on again after brewing is completed.

Brewing element
Heats water in a small well in the bottom of the pot; the steam pressure created forces hot water up the steam tube into the basket.

Thermostat
Switches off power to the brewing element at the end of the brewing cycle.

Power cord receptacle
The power cord plug fits over the pins; unplug the cord from the wall before unplugging it from the percolator to avoid damaging the pins.

Base ring

Selector switch
Adjusts the temperature setting of the thermostat.

Base plate

ACCESSING AND SERVICING THE PERCOLATOR COMPONENTS

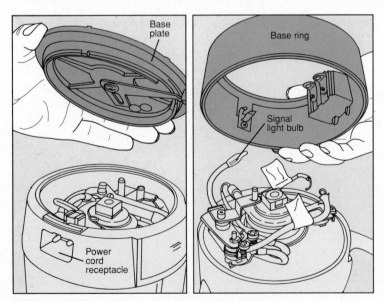

Base plate

Base ring

Signal light bulb

Power cord receptacle

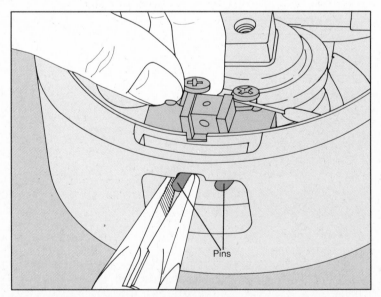

Pins

Disassembling the base. Turn off the percolator and unplug it from the wall outlet. Let it cool and empty it. Pull off the detachable power cord. Remove the lid, take out the basket and steam tube, and turn over the percolator. Remove the base-plate mounting screw and lift off the base plate *(above, left)*. Label and disconnect the wires to the power cord receptacle on the inside wall of the base ring. Gently pull the signal light bulb out of its clamp, then lift the ring off the bottom of the percolator *(above, right)*.

Servicing the power cord receptacle. Sand discolored pins with fine emery paper. If the pins are loose or damaged, disassemble the base *(left)*. To tighten loose pins, use long-nose pliers to turn their nuts clockwise while you hold the bracket terminals steady from behind *(above)*. To remove damaged pins and the flat washers beneath, turn the nuts counterclockwise. Buy identical replacements from an authorized service center. Install the pins and reassemble the percolator base. Cold check for leaking voltage *(page 114)*.

SERVICING THE PERCOLATOR

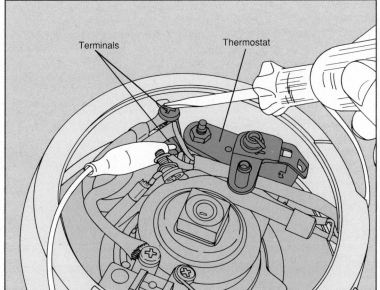

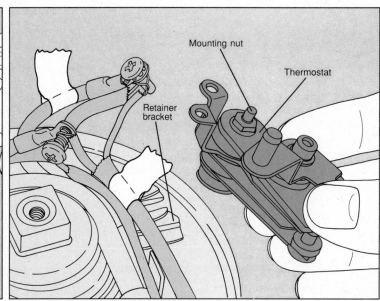

Servicing the thermostat. Wait until the percolator cools to room temperature, then disassemble the base *(page 46)*. Separate the thermostat contacts by hand and inspect them. If the contacts look pitted or black, slip a folded piece of fine emery paper between them and pull it back and forth gently, then follow with plain paper *(page 117)*. If the thermostat leaves are bent or broken, or the contacts do not close or are fused shut, replace the thermostat. To check the thermostat for less visible damage, test it. Label and disconnect the wires from one thermostat terminal. Touch a continuity tester probe

to each terminal *(above, left)*; the tester should light. To replace a faulty thermostat, label and disconnect the wires connected to the second thermostat terminal. Loosen the mounting nut on top of the thermostat assembly and slide the assembly sideways out of the retainer bracket *(above, right)*. Buy an identical replacement from an authorized service center. Install the new thermostat in the bracket and tighten the mounting nut. Reconnect the wires to the thermostat terminals. Reassemble the percolator by reinstalling the base ring and base plate, and cold check for leaking voltage *(page 114)*.

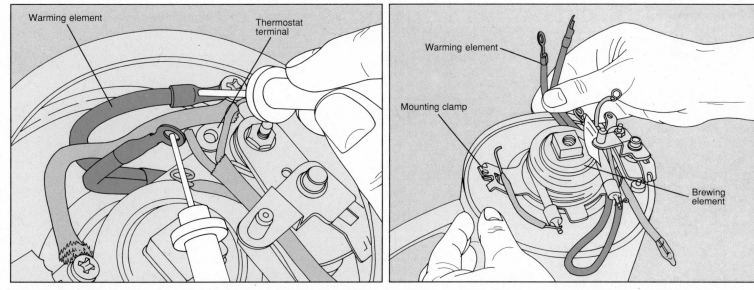

Testing and replacing the warming element. Unplug and turn off the percolator, and disassemble the base *(page 46)*. The warming element in the model shown here is an insulated wire coiled around the brewing element; label and disconnect one of its ends from a thermostat terminal. (You may also have to label and disconnect other wires from the same terminal in order to disconnect the warming element wire.) Set a multitester to test resistance *(page 113)*. Touch one probe to the free end of the warming element and touch the other probe to the other end where it connects to the other thermostat terminal *(above, left)*. The tester should show partial

resistance. If the warming element tests faulty, replace it. Disconnect the other warming-element wire end. (Label and disconnect all other wires attached to the second thermostat terminal, if necessary.) To remove the warming element, gently bend up the mounting clamp with a screwdriver to release the looped end of the element, then slide the element out from around the brewing element *(above, right)*. Buy a replacement warming element from an authorized service center and install it reversing the steps taken to remove the faulty one. Reassemble the percolator base and cold check for leaking voltage *(page 114)*.

HAIR DRYERS

A hand-held hair dryer works on the same principle as an electric heater; the heating elements' resistance to electrical current creates heat, and a motor operates a fan (or impeller) that blows air past the elements and out the nozzle. Instead of the universal or shaded pole motor found in most small appliances, the hair dryer uses a smaller, less powerful DC motor. One or more diodes wired to the switches reduce the voltage of household current and a series of motor diodes transforms it to direct current (DC). The typical hair dryer illustrated below has switches to control the on/off function, fan speed and heat level. Some hair dryers combine the on/off and fan speed controls in one switch.

Contrary to popular belief, many hair dryers can be serviced. Diagnose problems with the help of the Troubleshooting Guide at right; if replacement parts are needed to complete a repair, check small-appliance parts dealers for the right part for your model. As with many of the smaller household appliances, getting inside is half the battle. If your model is different from the one shown below, consult Tools & Techniques for tips on disassembly (page 115). However, a housing that is glued together is often an indication that the parts inside are not serviceable.

TROUBLESHOOTING GUIDE

SYMPTOM	PROCEDURE
Hair dryer doesn't work at all or works intermittently	Test power cord (p. 116) ▣○▲
	Service on/off switch (p. 49) ▣○
	Service elements (p. 50) ▣○
Hair dryer heats but fan works poorly or not at all	Clean hair dryer (p. 49) □○
	Service fan speed switch (p. 49) ▣○
Fan works but hair dryer doesn't heat	Test heat switch (p. 49) ▣○
	Service elements (p. 50) ▣○
Hair dryer works on only one heat setting	Service elements (p. 50) ▣○
	Service heat switch (p. 49) ▣○
	Service diode (p. 50) ▣○▲
Hair dryer works on only one fan setting	Service fan speed switch (p. 49) ▣○
	Service diode (p. 50) ▣○▲
Hair dryer overheats; may run a short time then stop or cool	Clean hair dryer (p. 49) □○
	Service elements (p. 50) ▣○

DEGREE OF DIFFICULTY: □ Easy ▣ Moderate ■ Complex
ESTIMATED TIME: ○ Less than 1 hour
▲ Special tool required

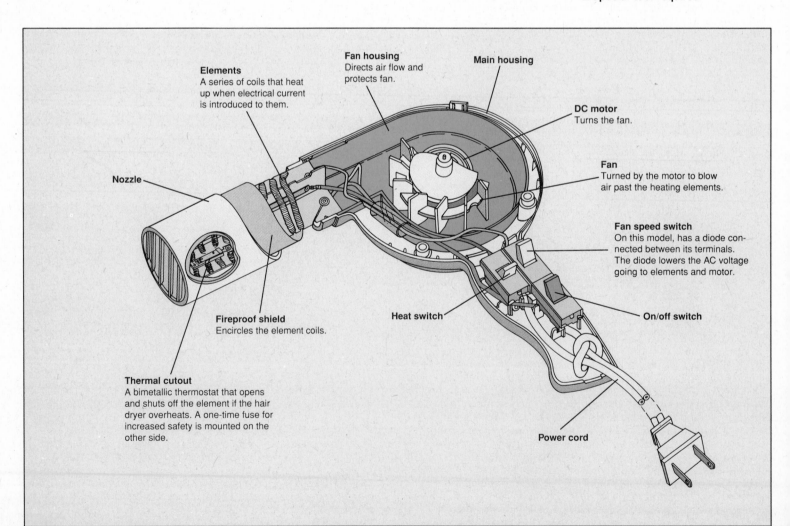

Elements
A series of coils that heat up when electrical current is introduced to them.

Fan housing
Directs air flow and protects fan.

Main housing

DC motor
Turns the fan.

Nozzle

Fan
Turned by the motor to blow air past the heating elements.

Fan speed switch
On this model, has a diode connected between its terminals. The diode lowers the AC voltage going to elements and motor.

Fireproof shield
Encircles the element coils.

Heat switch

On/off switch

Thermal cutout
A bimetallic thermostat that opens and shuts off the element if the hair dryer overheats. A one-time fuse for increased safety is mounted on the other side.

Power cord

CLEANING THE HAIR DRYER

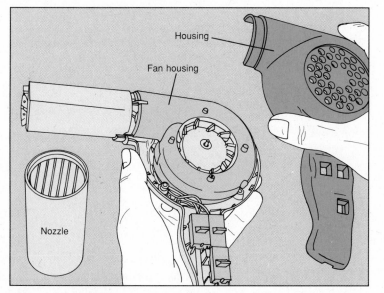

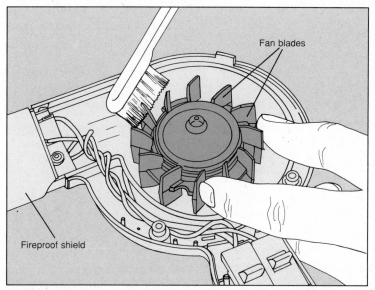

1 **Disassembling the hair dryer.** Turn off and unplug the hair dryer. On some models, the entire housing may be pulled apart in two pieces after you remove screws or pry open hidden tabs *(page 115)*. On others, the nozzle must be pulled off before the housing can be taken apart. On the model shown here, first remove the screw that holds the nozzle to the housing and pull the nozzle off, then remove the three screws that hold the halves of the housing together. Hold the hair dryer with the control switches up and pull off the upper half *(above)*.

2 **Cleaning the internal parts.** Pull hair out of the nozzle screen. Lift off the fan housing and remove hair by hand or with an old toothbrush *(above)*. Rotate the fan; if it turns stiffly, check for hair wrapped around the fan shaft and untangle it. Slip the fireproof shield off the end of the element assembly and inspect the element assembly for dirt and hair. Clean it gently with the toothbrush, taking care not to break or dislocate any element coils. Reassemble the hair dryer and cold check for leaking voltage *(page 114)*.

SERVICING THE SWITCHES

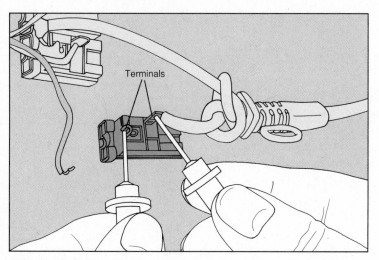

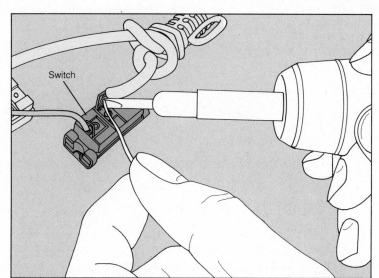

1 **Testing a switch.** Turn off and unplug the hair dryer and disassemble it *(step above)*. Most hair dryer switches, whether they control the on/off function, the heat or the fan, are simple switches with two terminals. To test one of these switches, lift it out of its seat in the handle, taking care not to tug on any wires. Disconnect one wire; on the model here, desolder it *(page 120)*. You may also have to desolder a diode wire if a diode is connected between the switch terminals. Use a continuity tester, or set a multitester to test continuity *(page 113)* and touch a tester probe to each terminal *(above)*. Flip the switch to each setting. The tester should show continuity in only one switch position. If the switch tests faulty, go to step 2 to replace it.

2 **Replacing a switch.** Label the wire from the other terminal and disconnect or desolder it *(page 120)*. If there is a diode connected to the switch terminal, note how its markings are oriented and disconnect it as well. Purchase an identical replacement switch from a small-appliance parts dealer. Using rosin-core solder and a soldering iron *(above)* solder *(page 121)* the wires, and the diode *(page 50, step 2)*, if there was one, to the switch terminals. Install the switch in the handle and reassemble the hair dryer. Cold check for leaking voltage *(page 114)*.

SERVICING THE DIODE

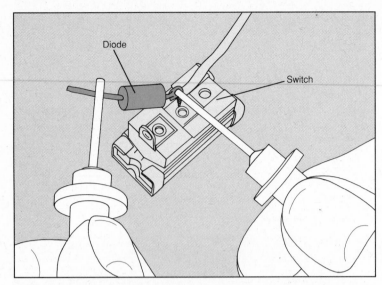

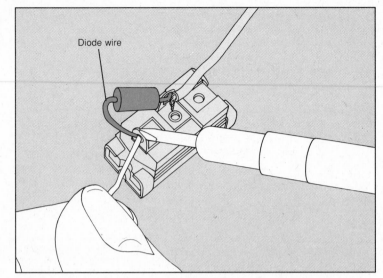

1 **Testing the diode.** Turn off and unplug the hair dryer and disassemble it *(page 49)*. Lift the switches out of the housing, taking care not to tug on any wires. Locate the diode connected to the terminals of one switch. Pull off or desolder *(page 120)* one diode wire. Set a multitester to its highest ohms setting to test resistance *(page 113)*. Touch a tester probe to each diode wire *(above)*, then reverse the probe positions and repeat the test. In one test, the multitester should register partial resistance. In the other, it should register infinite resistance. If the diode tests faulty, go to step 2 to replace it.

2 **Replacing the diode.** Note the markings on the diode and make a sketch of how the diode is oriented so you can install the replacement diode the same way. To remove the faulty diode, pull off or desolder *(page 120)* its other wire. Take the diode to a small-appliance parts dealer or electronics parts supplier and buy an exact replacement. Consult the sketch and install the diode, soldering *(page 121)* the wires to their correct terminals *(above)*. Apply heat quickly and carefully to avoid damaging the diode. Reinstall the switch in the housing and reassemble the hair dryer. Cold check for leaking voltage *(page 114)*.

SERVICING THE ELEMENTS

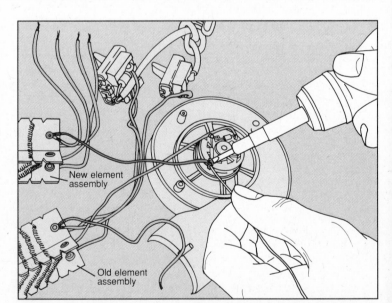

1 **Inspecting the element assembly.** Turn off and unplug the hair dryer, and disassemble it *(page 49)*. Slip off the fireproof shield *(above)*. Inspect the thermal cutout located under the coils. If its bimetal leaves look discolored, use a folded piece of fine emery paper to polish them *(inset)*, then clean them with plain paper. Check that the element coils are not touching each other and that their ends are firmly connected to their terminals. Examine the coils for evidence of burning or breakage. If the coils are damaged or the cutout contacts are fused or broken, replace the element assembly *(step 2)*.

2 **Replacing the element assembly.** Take the hair dryer to a small-appliance parts dealer and buy an exact replacement element assembly. Lay the new assembly alongside the old one in the same orientation. Using desoldering braid, desolder *(page 120)* one of the old element assembly wires, then use rosin-core solder to solder *(page 121)* the corresponding wire of the new element assembly in place *(above)*. Work from wire to wire until all the new wires are soldered to their correct terminals. Reassemble the hair dryer and cold check for leaking voltage *(page 114)*.

HEATERS

Most modern electric space heaters, regardless of their design, use the same basic components to turn electricity to heat: A power cord, one or more heating elements and a control thermostat. Some, such as the popular fan-forced radiant heater shown on page 52, include a fan and a shiny reflector to help disperse the heat. Since this type heats the air directly with exposed heating elements, it also has safety devices such as a tip-over switch, an overheat protector and a warning buzzer. In contrast, the heating elements on the oil-filled heater illustrated on page 56 are immersed in diathermic oil. When the oil has reached a set temperature, the thermostat turns off the elements, leaving the oil to heat the room through the heater's fins.

If your heater doesn't work efficiently and quietly, consult the Troubleshooting Guide below. Always turn off and unplug the heater before inspecting it. During repair, pay close attention to the orientation of the parts; a missing or incorrectly placed part could be a fire or shock hazard. Be on the lookout for burned wires, broken terminals or damaged elements, and review Tools & Techniques *(page 110)* for helpful repair tips.

Follow the use and care recommendations in your owner's manual. Place heaters with exposed elements away from furniture and draperies. When used around small children, install a protective screen *(page 11)* to prevent small hands from reaching through the heater grille. A dirty heater is not only a fire hazard but will heat less efficiently; vacuum air vents regularly and clean reflectors with a vacuum-cleaner crevice attachment or a soft cloth.

TROUBLESHOOTING GUIDE

SYMPTOM	POSSIBLE CAUSE	PROCEDURE
ALL HEATERS		
Heater doesn't work at all	Heater unplugged or turned off	Plug in and turn on heater
	No power to outlet or outlet faulty	Reset breaker or replace fuse *(p. 112)* □○; have outlet serviced
	Power cord faulty	Test and replace power cord *(p. 116)* ◼○▲
	Wire connection loose or broken	Repair wire connection *(p. 118)* ◼○
	Control thermostat faulty	Test and replace control thermostat *(fan-forced radiant, p. 53; oil-filled, p. 56)* ◼○
	Wattage switch faulty	Test and replace wattage switch *(fan-forced radiant, p. 53; oil-filled, p. 57)* ◼○▲
	Heating elements faulty	Test and replace heating elements *(fan-forced radiant, p. 55; oil-filled, p. 57)* ◼◕▲
Heater doesn't heat on some wattage settings	Wattage switch faulty	Test and replace wattage switch *(fan-forced radiant, p. 53; oil-filled, p. 57)* ◼○▲
	Heating elements faulty	Test and replace heating elements *(fan-forced radiant, p. 55; oil-filled, p. 57)* ◼◕▲
Heater runs continuously; doesn't cycle on and off	Cold draft on heater	Move heater out of draft; don't place below open window
	Control thermostat faulty	Test and replace control thermostat *(fan-forced radiant, p. 53; oil-filled, p. 56)* ◼○
Heater blows circuit breaker or fuse repeatedly	Household electrical circuit overloaded	Reset breaker or replace fuse *(p. 112)* □○, reduce number of appliances on circuit
Heater gives electrical shock	Internal wiring or component grounded to chassis	Cold check for leaking voltage *(p. 114)* ◼○▲
FAN-FORCED RADIANT HEATERS ONLY		
Heater doesn't heat	Heater overheated	Turn off heater, let cool for 30 minutes; place in an open area
	Overheat protector faulty	Test and replace overheat protector *(p. 54)* ◼○
Heater turns on, then stops and doesn't cycle back on	Control thermostat faulty	Test and replace control thermostat *(p. 53)* ◼○
	Fan motor faulty	Test and replace fan motor *(p. 54)* ◼○▲
	Air intake or outlet vents blocked	Remove obstructions, vacuum vents
Heater heats, but fan doesn't spin or spins unevenly	Fan loose, bent or obstructed	Tighten or replace fan *(p. 54)* ◼○
	Fan motor faulty	Test and replace fan motor *(p. 54)* ◼○▲
Heater is noisy	Fan loose, bent or obstructed	Tighten or replace fan *(p. 54)* ◼○
	Motor bearings dry or worn	Lubricate motor bearings *(p. 125)* ◼○

DEGREE OF DIFFICULTY: □ Easy ◼ Moderate ◼ Complex
ESTIMATED TIME: ○ Less than 1 hour ◕ 1 to 3 hours ● Over 3 hours ▲ Special tool required

FAN-FORCED RADIANT HEATERS

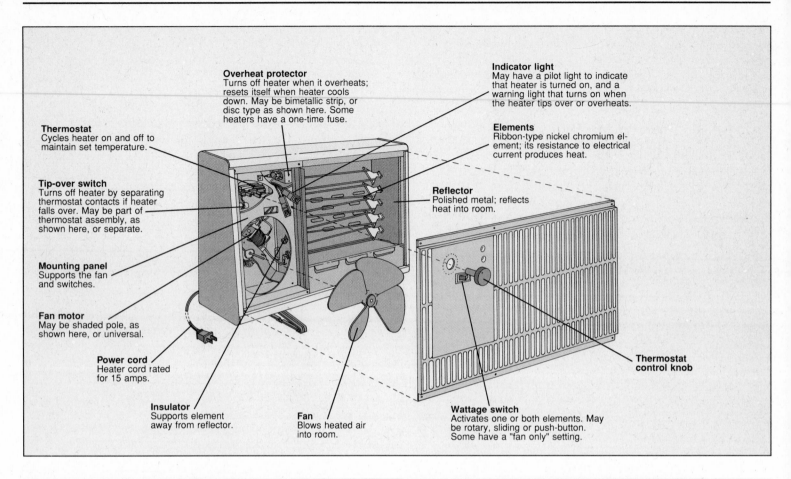

Overheat protector
Turns off heater when it overheats; resets itself when heater cools down. May be bimetallic strip, or disc type as shown here. Some heaters have a one-time fuse.

Indicator light
May have a pilot light to indicate that heater is turned on, and a warning light that turns on when the heater tips over or overheats.

Thermostat
Cycles heater on and off to maintain set temperature.

Elements
Ribbon-type nickel chromium element; its resistance to electrical current produces heat.

Tip-over switch
Turns off heater by separating thermostat contacts if heater falls over. May be part of thermostat assembly, as shown here, or separate.

Reflector
Polished metal; reflects heat into room.

Mounting panel
Supports the fan and switches.

Fan motor
May be shaded pole, as shown here, or universal.

Power cord
Heater cord rated for 15 amps.

Thermostat control knob

Insulator
Supports element away from reflector.

Fan
Blows heated air into room.

Wattage switch
Activates one or both elements. May be rotary, sliding or push-button. Some have a "fan only" setting.

ACCESS TO INTERNAL PARTS

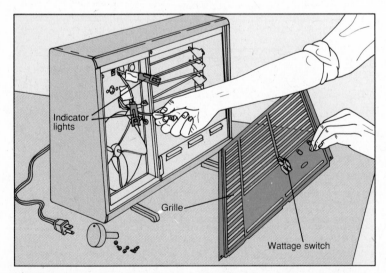

Indicator lights

Grille

Wattage switch

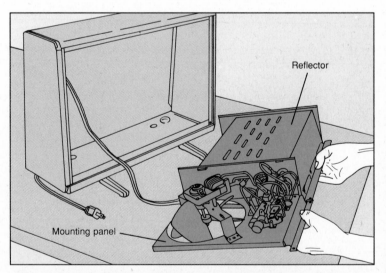

Reflector

Mounting panel

1 **Removing the grille.** Unplug the heater and let it cool. Pull off the control knob. On some models you may need first to remove a setscrew that holds the knob to the shaft. Remove the screws around the edge of the grille. Lift the grille away from the chassis just enough to reach the wires connected to the wattage switch. Label the wires and pull them, and the indicator lights, off the back of the grille *(above)*. You can now test and replace the wattage switch.

2 **Removing the mounting panel and reflector.** Locate the strain relief grommet where the power cord enters the back of the heater, and remove it *(page 116, step 3)*. Grasp the mounting panel and reflector and pull them out of the cabinet *(above)*. You can now service the control thermostat assembly, fan, fan motor, heating elements and overheat protector.

TESTING AND REPLACING THE CONTROL THERMOSTAT ASSEMBLY

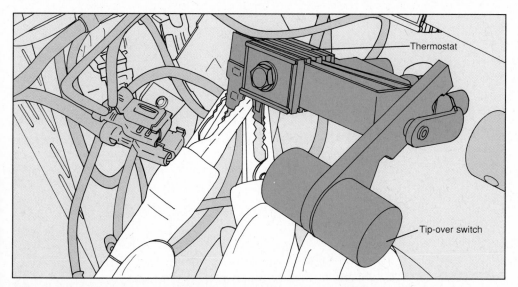

Thermostat

Tip-over switch

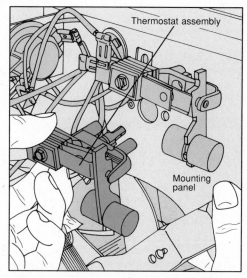

Thermostat assembly

Mounting panel

1 Testing the thermostat and tip-over switch. Unplug the heater and let it cool. Access the heater's internal parts *(page 52)*. Locate the thermostat and inspect it. If the contacts are dirty, clean them *(page 117)*. If the terminals look burned, replace the thermostat *(step 2)*. To check for less visible damage, use a continuity tester, or set a multitester to test continuity *(page 113)*. If your unit has a tip-over switch, place the heater in an upright position. Pull one wire connector off the thermostat and clip a tester probe to each terminal. Rotate the control shaft to its highest setting; the tester should indicate continuity. Rotate the shaft to its lowest setting; the tester should indicate no continuity. If the thermostat has a tip-over switch, rotate the control shaft to its highest setting. Swing the tip-over switch arm toward you and then away from you. In both positions, the tester should indicate no continuity. If the tip-over switch is jammed or faulty, or the thermostat fails any test, replace the thermostat assembly.

2 Replacing the thermostat assembly. Label and disconnect the wires. Unscrew the thermostat assembly from the mounting panel *(above)*. Buy an exact replacement from an electrical parts supplier or the appliance manufacturer; if the old one had a tip-over switch, make sure the new one does, too. Screw the new thermostat assembly to the mounting panel. Reconnect the wires and remove the labels. Reassemble the heater, reversing the steps you took to disassemble it, and cold check for leaking voltage *(page 114)*.

SERVICING THE WATTAGE SWITCH

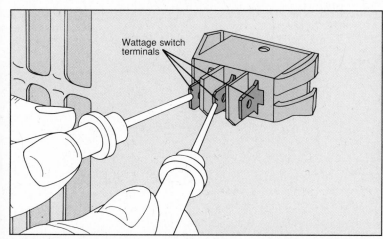

Wattage switch terminals

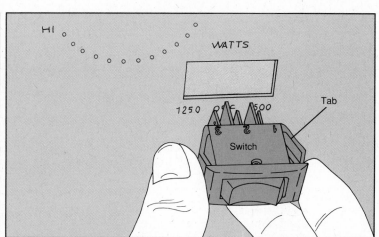

HI · WATTS · 1250 · 1500 · Tab · Switch

Testing and replacing the wattage switch. Unplug the heater and let it cool. Remove the grille *(page 52, step 1)*. Examine the switch; if the terminals are damaged or burned, replace the switch. To check for less visible damage, use a continuity tester or set a multitester to test continuity *(page 113)*. A three-setting wattage switch (high/low/off), such as the one shown here, has three terminals. Set the switch to the OFF position and touch the multitester probes to all three possible pairings of terminals *(above, left)*; the multitester should register no continuity for each pair. Then set the switch to one of the two wattage settings and repeat the test between all three possible pairings of terminals; the multitester should indicate continuity once and only once. Repeat this test with the switch set to the other wattage setting; again the multitester should show continuity only once. If the switch fails any test, replace it. Remove the old switch by squeezing the tabs on its sides and pushing it out the front of the mounting panel. Buy an identical switch from an electrical parts supplier or the manufacturer. Push the switch in through the front of the panel with the numbers facing up *(above, right)*. Reconnect the wires and remove the labels. Reassemble the heater and cold check for leaking voltage *(page 114)*.

TESTING AND REPLACING THE OVERHEAT PROTECTOR

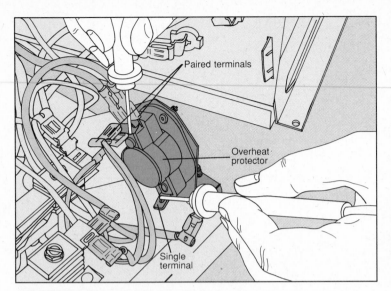

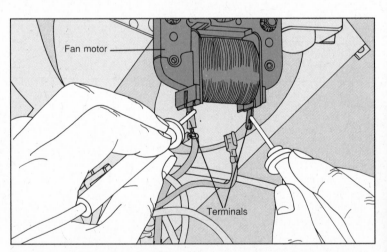

1 **Testing the overheat protector.** Unplug the heater and let it cool. Access the heater's internal parts *(page 52)*. Locate the overheat protector, near the element or behind the reflector. It may be a bimetal strip type, resembling a thermostat, or a disc type, as shown here. Test the bimetal strip type as you would a thermostat *(page 53, step 1)*. To test a disc type with three terminals, two of which are paired, disconnect the wire from the single terminal. Use a continuity tester, or set a multitester to test continuity *(page 113)*. Touch one probe to the free terminal and the other probe to either of the paired terminals *(above)*. The tester should indicate continuity. If the protector tests faulty, replace it.

2 **Replacing the overheat protector.** Label and disconnect the wires. Unscrew the protector from its mounting bracket. It may be easier to remove the bracket first *(above)* and then unscrew the protector from the bracket. Buy an identical replacement from an electrical parts supplier or the appliance manufacturer. Install the overheat protector, reversing the steps you took to remove the faulty one. Reconnect the wires and remove the labels. Reinstall the mounting panel, reflector and grille and cold check for leaking voltage *(page 114)*.

SERVICING THE FAN AND FAN MOTOR

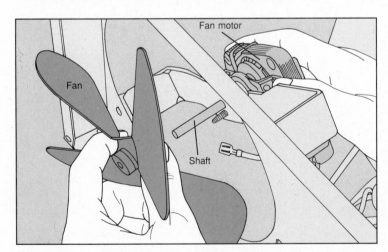

Inspecting and replacing the fan. Unplug the heater and let it cool. Remove the grille *(page 52, step 1)* and check for anything hindering the fan's spinning. Tighten a loose fan by pushing it firmly onto its shaft or by tightening its setscrew. If the fan doesn't spin easily, the fan motor may need cleaning *(page 125)*. If the blades are bent or broken, replace the fan. Remove the mounting panel and reflector *(page 52, step 2)*. Pull the fan off its shaft *(above)*; first remove its setscrew or retaining clip, if any. Buy an exact replacement from an electrical parts supplier or the manufacturer. Push the new fan onto the shaft and reinstall any fasteners. Reassemble the heater and cold check for leaking voltage *(page 114)*.

Testing and replacing the fan motor. Unplug the heater and let it cool. Access its internal parts *(page 52)*. Disconnect one wire from the motor. Set a multitester to RX1 *(page 113)* and touch a tester probe to each motor terminal *(above)*. The tester should show partial resistance. If the motor tests faulty, replace it. Label and pull off the other wire. Unscrew and remove the fan motor from its bracket, keeping track of all screws, washers and spacers. Pull off the fan *(left)*. Buy an exact replacement fan motor from an electrical parts supplier. Position the spacers over the holes in the bracket and carefully screw on the new motor. Reconnect the wires and remove the labels. Reinstall the fan *(left)* and reassemble the heater. Cold check for leaking voltage *(page 114)*.

TESTING AND REPLACING THE HEATING ELEMENTS

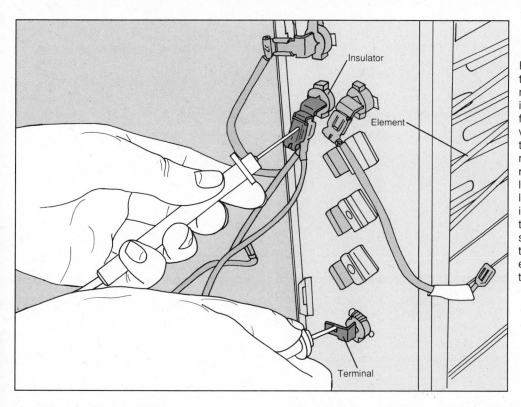

1 **Inspecting and testing the elements.** Unplug the heater and let it cool. Access the heater's internal parts *(page 52)*. Repair any loose or damaged wire connections *(page 118)*. Inspect the elements; the model shown here has two of them. If either is broken, burned-looking or touching the reflector, replace it *(step 2)*. To test for less visible damage, set a multitester to test resistance *(page 113)*. Locate the element terminals and their insulators on the outside of the reflector and determine which terminals belong to which elements. To test one element, label and disconnect the wires from one of its terminals and touch a multitester probe to each element terminal *(left)*. The tester should indicate low resistance. Repeat the test on the other element's terminals. If either element shows infinite resistance or no resistance, replace it *(step 2)*.

2 **Replacing a heating element.** Label and disconnect any remaining wires. Sketch the element's mounting pattern for correct installation. On some models, the element and reflector are replaced as an assembly. On other models, elements may be simply unscrewed and lifted out. On the model shown here, the ribbon elements are secured with twist-on insulated terminals. To release one end of an element, rotate its insulated terminal one-quarter turn, then push it through its slot in the reflector *(above, left)*. On older models of this type, you may have to use a screwdriver carefully to pry open the insulators' retaining tabs *(inset)*. After releasing one terminal, unwind the element from around the large Y-shaped insulators *(above, right)*. Then release the element's second insulated terminal, and lift out the element. Buy an exact replacement element from an electrical parts supplier or the manufacturer. Insert the new element's insulated terminal into its slot from inside the reflector, and lock it in place from the outside by rotating it one-quarter turn. Mount the element on the Y-shaped insulators, following the sketch you made. Have a helper keep the mounted portions as taut as possible as you wrap the element around the insulators, without stretching it. Secure the second terminal as you did the first. Check that the tension is even along the element. Connect the wires to the element terminals and remove the labels. Reassemble the heater and cold check for leaking voltage *(page 114)*.

OIL-FILLED HEATERS

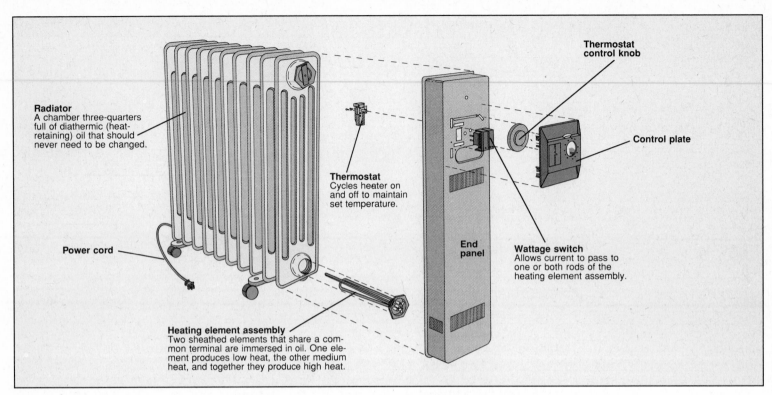

Radiator
A chamber three-quarters full of diathermic (heat-retaining) oil that should never need to be changed.

Thermostat
Cycles heater on and off to maintain set temperature.

Power cord

Heating element assembly
Two sheathed elements that share a common terminal are immersed in oil. One element produces low heat, the other medium heat, and together they produce high heat.

Thermostat control knob

Control plate

End panel

Wattage switch
Allows current to pass to one or both rods of the heating element assembly.

Removing the end panel to access internal parts. Unplug the heater and let it cool. Locate the end-panel mounting screws and remove them. On the model shown above, the screw is above the control plate. Lift away the end panel, being careful not to pull off any wire connections. You now have access to the thermostat, wattage switch, heating element terminals and power cord terminals.

SERVICING THE THERMOSTAT

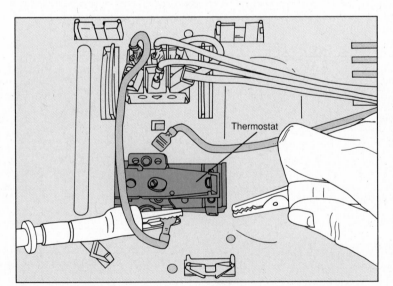

Thermostat

1 Testing the thermostat. Unplug the heater and let it cool. Remove the end panel *(step above)*. Use a continuity tester, or set a multitester to test continuity *(page 113)*. Rotate the control knob to the OFF position. Pull a wire connector off one of the thermostat terminals and clip a probe to each terminal *(above)*. The tester should indicate no continuity. Rotate the knob toward its highest setting, and listen for a click. After the click, the tester should show continuity. If the thermostat fails either test, replace it.

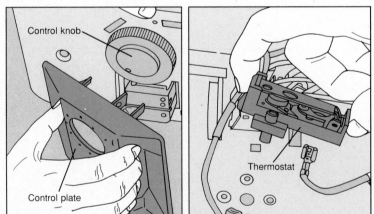

Control knob

Control plate

Thermostat

2 Replacing the thermostat. Label and disconnect the remaining wire. Work the control plate tabs free of their slots in the end panel, turn over the panel and pull off the plate *(above, left)*. To access the thermostat mounting screws, pull off the control knob and spacer. Unscrew the thermostat and remove it. Buy an exact replacement from an electrical parts supplier or the manufacturer. Position the new thermostat on the inside of the end panel *(above, right)* and resecure the mounting screws. Then reposition the spacer and control knob on the thermostat shaft, and reinstall the control plate. Reconnect the wires and remove the labels. Screw the end panel onto the radiator and cold check the heater for leaking voltage *(page 114)*.

TESTING AND REPLACING THE WATTAGE SWITCH

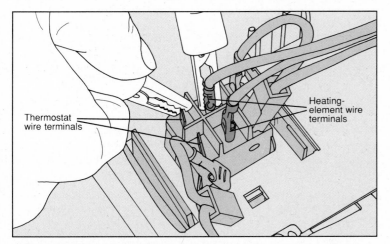

Thermostat wire terminals

Heating-element wire terminals

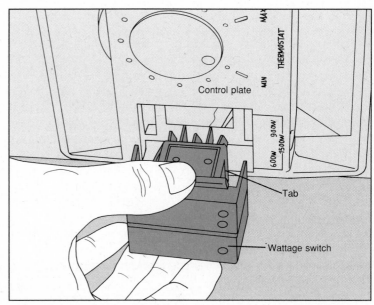

Control plate

Tab

Wattage switch

1 Testing the wattage switch. Unplug the heater and let it cool. Remove the end panel *(page 56)*. Locate the wattage switch terminals; if they look burned or damaged, replace the switch *(step 2)*. The wattage switch shown here has three pairs of terminals; two terminals are connected to the thermostat, two to the heating elements and two to the power source. To test it, set a multitester to test continuity *(page 113)*. Disconnect the thermostat wires from the switch. Clip a tester probe to one of the free terminals and to the heating-element wire terminal next to it *(above)*. Set the switch to all three wattage settings in turn; the multitester should indicate continuity in two of the three settings. Repeat the test with the second thermostat wire terminal and the second heating-element wire terminal. There should be continuity in two of the three switch settings. If the switch fails either test, replace it.

2 Replacing the wattage switch. Label and disconnect the remaining switch wires. Remove the switch by squeezing its tabs together from inside the end panel and pushing it out through the opening in the control plate. Buy an identical replacement for your model from the manufacturer. To install it, push the switch into its slot in the control plate until it clicks into place *(above)*. Reconnect the wires and remove the labels. Reinstall the end panel and cold check for leaking voltage *(page 114)*.

SERVICING THE HEATING ELEMENTS

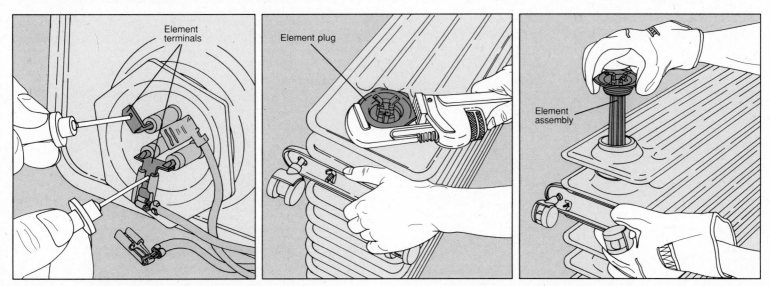

Element terminals

Element plug

Element assembly

Testing and replacing the elements. Unplug the heater and let it cool. Remove the end panel *(page 56)*. Find the common terminal shared by the two elements. To test an element, pull a wire off its non-common terminal. Set a multitester to RX1 *(page 113)* and touch one probe to the free terminal and one to the common terminal *(above, left)*. The tester should indicate partial resistance. Reconnect the wire and pull the wire off the other element's non-common terminal. Repeat the test. If either element fails this test, replace the assembly. Label and disconnect the wires. Lay the

heater on its end on newspaper, with the element terminals facing up. Use a pipe wrench to loosen the element plug counterclockwise *(above, center)*. Lift the element assembly out of the chamber *(above, right)*, taking care not to let any oil touch your hands. Dispose of the elements in a plastic bag. Buy a replacement assembly from the manufacturer. Lower the assembly into the chamber and tighten the plug clockwise. Wipe off spilled oil and stand the heater upright. Reconnect the wires and remove the labels. Reinstall the end panel, and cold check for leaking voltage *(page 114)*.

FANS

The cooling breeze of a portable electric fan is an indispensable and economical comfort on a hot, still day. Its simple design not only makes a fan reliable but when its parts eventually wear out, fixing the fan is simple as well. Most small fans use a shaded-pole or split-phase induction motor to spin the propeller-like blades, and have a push-button or rotary multispeed switch for selecting fan speeds. If the soft whirr of your fan's blades is replaced by rattling or grinding, or worse, if the blades stop spinning altogether, consult the Troubleshooting Guide at right and correct the problem as described in this chapter. Most replacement parts are inexpensive, but if you suspect a faulty motor, it may be cheaper to replace the fan.

A typical portable oscillating fan is pictured below. The oscillating fan has a gear assembly, also run by the motor, that swivels the fan back and forth. The gears are frequently made of plastic. The primary gear, in particular, is likely to have its teeth stripped before any other fan parts show wear. Replacing the primary gear is easy, but if the less accessible secondary gear is stripped, take the fan for professional service.

When disassembling the fan, check for a capacitor—it looks like a battery. Discharge it according to the instructions in Tools & Techniques *(page 114)*. Keep track of any variations in disassembly steps for your model and always replace parts with identical replacements.

TROUBLESHOOTING GUIDE

SYMPTOM	PROCEDURE
Fan doesn't run at all	Reset circuit breaker or replace fuse *(p. 112)* □○; have outlet serviced
	Test power cord *(p. 116)* ▣○▲
	Clean speed control switch *(p. 117)* □○ and test it *(p. 59)* □○
	Repair wire connections *(p. 118)* □○
Motor hums but fan blade doesn't turn	Service capacitor *(p. 60)* □○▲
	Take fan for professional service
Fan works on only some speeds	Service speed control switch *(p. 59)* □○
	Repair wire connections *(p. 118)* □○
Fan is noisy or vibrates	Tighten loose clips and screws
	Level oscillating fan on padded surface
	Secure fan blades *(p. 59)* □○
Oscillating fan jerks when it swivels	Service gear assembly *(p. 60)* □○
Fan blades turn sluggishly	Service fan blades *(p. 59)* □○; lubricate bearings according to owner's manual
Fan gives electrical shock	Cold check for leaking voltage *(p. 114)* □○; repair wire connections *(p. 118)* □○

DEGREE OF DIFFICULTY: □ Easy ▣ Moderate ■ Complex
ESTIMATED TIME: ○ Less than 1 hour
▲ Special tool required

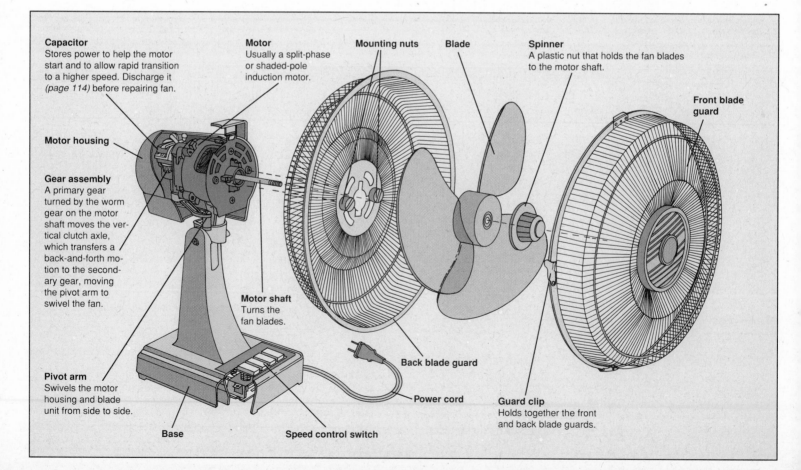

Capacitor
Stores power to help the motor start and to allow rapid transition to a higher speed. Discharge it *(page 114)* before repairing fan.

Motor housing

Gear assembly
A primary gear turned by the worm gear on the motor shaft moves the vertical clutch axle, which transfers a back-and-forth motion to the secondary gear, moving the pivot arm to swivel the fan.

Pivot arm
Swivels the motor housing and blade unit from side to side.

Motor
Usually a split-phase or shaded-pole induction motor.

Motor shaft
Turns the fan blades.

Base

Mounting nuts

Blade

Spinner
A plastic nut that holds the fan blades to the motor shaft.

Front blade guard

Back blade guard

Power cord

Guard clip
Holds together the front and back blade guards.

Speed control switch

SERVICING THE FAN BLADES

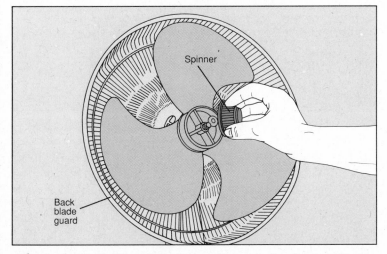

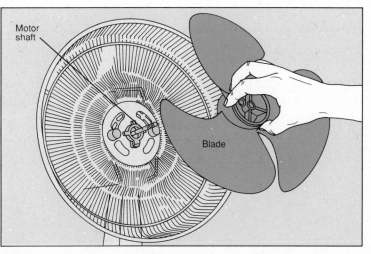

1 Accessing the fan blades. Turn off and unplug the fan. To access the blades on an oscillating fan, pry up the clips that hold the front blade guard to the back one and lift off the guard. To access a box fan blade unit, simply unscrew the front grille from the fan frame and lift it off. Free a plastic blade unit on a typical oscillating fan by unscrewing the spinner nut clockwise *(above)*; use pliers to loosen a stiff nut. If the unit is metal, free it by loosening the setscrew that holds it to the motor shaft.

2 Removing the fan blades. Slide the blade unit off the end of the motor shaft *(above)*. If the blades are dirty, wipe them with a slightly damp cloth. Inspect the blades for cracks or warps that may cause noise and vibration, and replace them if worn or damaged. Take the fan to an appliance parts dealer when you buy a replacement blade unit and make sure that the new one fits correctly. Also buy a replacement spinner for a fan with plastic blades. Install the unit and reassemble the fan.

SERVICING THE SPEED CONTROL SWITCH

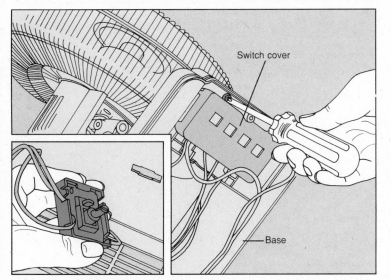

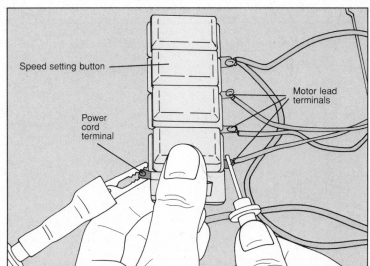

1 Accessing the speed control switch. To gain access to an oscillating fan speed control switch, turn off and unplug the fan and lay it on its side. Unscrew and remove the base plate. Unscrew the switch cover *(above)* and lift it off the switch assembly. Remove the switch-assembly mounting screws and lift the assembly out of the base, taking care not to tug on any wires. To test the switch, go to step 2. To gain access to a box fan speed control switch, unscrew and remove the front grille from the fan frame. Pull the dial or knob off the top of the switch, unscrew or unclip the switch from the fan frame and pull it out *(inset)*. Test it like a humidifier speed control switch *(page 89)*.

2 Testing and replacing the speed control switch. To test the push-button switch on an oscillating fan, use a continuity tester, or set a multitester to test continuity *(page 113)*. Pull off or desolder *(page 120)* the power cord wire from the terminal opposite the motor lead terminals. Clip a tester probe to the terminal, lock down the first speed setting button and touch the other probe to the terminal nearest that button *(above)*. The tester should show continuity. Repeat the test for each speed setting. If the switch tests faulty, label and pull off or desolder *(page 120)* the other wires and remove the switch. Buy an exact replacement and install it, reversing the steps taken for removal. Reassemble the fan and cold check for leaking voltage *(page 114)*.

SERVICING THE CAPACITOR

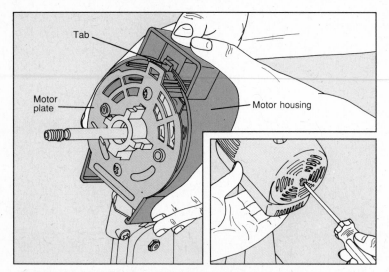

1 **Removing the motor housing.** Before removing the motor housing, make a capacitor discharging tool *(page 114)*. Remove the fan blades *(page 59)*, then unscrew and remove the mounting nuts that hold the back blade guard to the motor housing and lift the guard off the housing. Locate the screws or tabs that hold the motor housing together. On the fan shown here, first unscrew the retaining screw from the back of the housing *(inset)*. Then depress the tab at the top of the housing and unhook it from the motor plate *(above)*. Pull the housing back off the motor assembly. Before proceeding, locate the battery-like capacitor mounted near the motor and discharge it *(page 114)*.

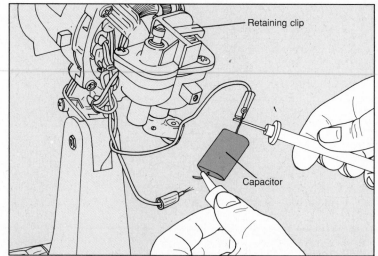

2 **Testing and replacing the capacitor.** Slide the capacitor out of its retaining clip behind the motor. If the terminal wires or connectors are burned, replace the capacitor. To test for less visible damage, set a multitester to RX1 *(page 113)*. Desolder *(page 120)* one capacitor terminal wire. Touch a tester probe to each capacitor terminal wire *(above)*. The tester needle should rise, then fall. If the needle swings to the right and stays, or if it doesn't move at all, the capacitor is faulty. Desolder *(page 120)* the other terminal wire and take out the capacitor. Buy an exact replacement from an authorized service center. Solder *(page 121)* the new capacitor terminal wires and install the capacitor. Reassemble the fan and cold check for leaking voltage *(page 114)*.

SERVICING THE GEARS

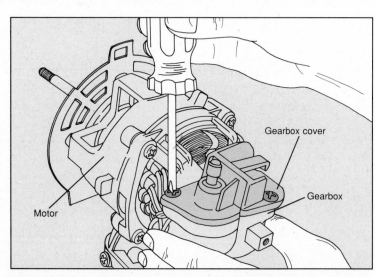

1 **Accessing the gears.** Turn off and unplug the fan. Make a capacitor discharging tool *(page 114)*. Remove the blade guard and the motor housing *(step above)*. Locate the battery-like capacitor mounted near the motor, discharge it *(page 114)* and unclip it from the retaining clip. To gain access to the gears on the model shown here, unscrew the gearbox cover behind the motor *(above)* and lift it off. On some models, you may first have to remove a metal counterweight; on others, the gears may be accessible once the motor housing is removed.

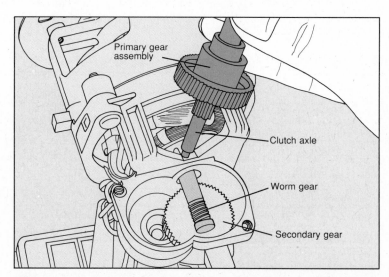

2 **Inspecting and replacing the gear assembly.** Lift out the primary gear assembly *(above)*. If the teeth on the clutch axle or gear are broken or stripped, install an exact replacement gear assembly from an authorized service center. If the teeth on the secondary gear are damaged, take the fan for service. Clean the worm gear with a small brush, and wipe off old lubricant and debris inside the gearbox. Lubricate the teeth of the new gear assembly and the top of the secondary gear with a high-temperature multipurpose grease. Seat the gear assembly and lubricate its top. Reassemble the fan and cold check for leaking voltage *(page 114)*.

SHAVERS

The electric shaver with its replaceable cutting blades is a far cry from the old-fashioned straight razor and strop, but the technology that makes it such a blessing for busy people also makes it more likely to need repair. Consult the Troubleshooting Guide below to help you diagnose shaver symptoms. Before disassembling a shaver, check whether it is still under warranty; if so, take it to an authorized service center.

Most shaver malfunctions are experienced as an uneven shave, or a snagging or pinching as you run the shaver head over your skin. When this happens, consult the owner's manual and this chapter for tips on cleaning and maintaining the cutter assembly. A chief difference among electric shavers lies in the design of their cutter assemblies. The foil head shaver illustrated on page 62 has a single flexible screen that covers rows of vibrating cutters. The rotary head shaver shown on page 64 has circular screens, or combs, in the shaver head, each covering a small rotating cutter.

Most modern shavers are now rechargeable. Their motors run on direct current (DC) stored by nickel-cadmium (nicad) batteries in the shaver body. A shaver that can also operate when plugged in has a small transformer that converts alternating current (AC) to DC. Shaver motors can be replaced inexpensively. Replace faulty batteries in a rechargeable shaver as described in Tools & Techniques *(page 122)*.

TROUBLESHOOTING GUIDE

SYMPTOM	POSSIBLE CAUSE	PROCEDURE
RECHARGEABLE SHAVERS (FOIL HEAD AND ROTARY HEAD)		
Shaver doesn't run at all	No power to outlet or outlet faulty	Reset breaker or replace fuse *(p. 112)* □○; have outlet serviced
	Nicad batteries need charging	Consult owner's manual; charge according to instructions
	Power cord faulty	Test and replace power cord *(p. 116)* ■○▲
	On/off switch faulty	Service on/off switch *(foil head shavers, p. 63* ■○; *rotary head shavers, p. 66)* ■◑▲
	Motor faulty	Test and replace DC motor *(foil head shavers, p. 64; rotary head shavers, p. 67)* ■○▲
	Nicad batteries or charging circuit faulty	Test charging circuit and replace nicad batteries *(p. 122)* ■○▲; or take shaver for professional service
Shaver needs frequent recharging	Nicad batteries or charging circuit faulty	Test charging circuit and replace nicad batteries *(p. 122)* ■○▲; or take shaver for professional service
Shaver runs sluggishly	One nicad battery faulty	Test charging circuit and replace nicad batteries *(p. 122)* ■○▲; or take shaver for professional service
NONRECHARGEABLE SHAVERS (FOIL HEAD AND ROTARY HEAD)		
Shaver doesn't run at all	No power to outlet or outlet faulty	Reset breaker or replace fuse *(p. 112)* □○; have outlet serviced
	Power cord faulty	Test and replace power cord *(p. 116)* ■○▲
	On/off switch faulty	Test switch *(p. 117)* ■○▲; take shaver for professional service
	Motor faulty	Service universal motor *(rotary head shaver, p.123)* ■○▲; take shaver with vibrator motor for professional service *(foil head shaver)*
FOIL HEAD SHAVERS		
Shaver snags and pinches	Screen corroded or torn	Replace screen *(p. 62)* □○
Shaver shaves unevenly	Cutter blades or head assembly damaged	Replace cutter or head assembly *(p. 62)* □○
Trimmer is noisy	Trimmer blades need lubrication	Lubricate trimmer *(p. 62)* □○
Trimmer snags or cuts unevenly	Trimmer blades dull or corroded	Replace head assembly *(p. 62)* □○
Motor runs but shaver doesn't work or works noisily	Oscillators worn	Replace oscillators *(p. 63)* □○
ROTARY HEAD SHAVERS		
Shaver shaves poorly; heads slow down while shaving	Combs and cutters dry or worn	Spray combs with manufacturer-recommended lubricant while shaver is running; replace worn combs and cutters *(p. 65)* □○
Trimmer snags or cuts unevenly	Trimmer blades dull or corroded	Replace trimmer *(p. 66)* □○
Motor runs but trimmer doesn't work	Trimmer arm worn or broken	Replace trimmer arm *(p. 66)* □○
Shaver is noisy or vibrates	Gears worn or need lubrication	Service gears *(p. 65)* □○

DEGREE OF DIFFICULTY: □ Easy ■ Moderate ■ Complex
ESTIMATED TIME: ○ Less than 1 hour ◑ 1 to 3 hours ● Over 3 hours   ▲ Special tool required

FOIL HEAD SHAVERS

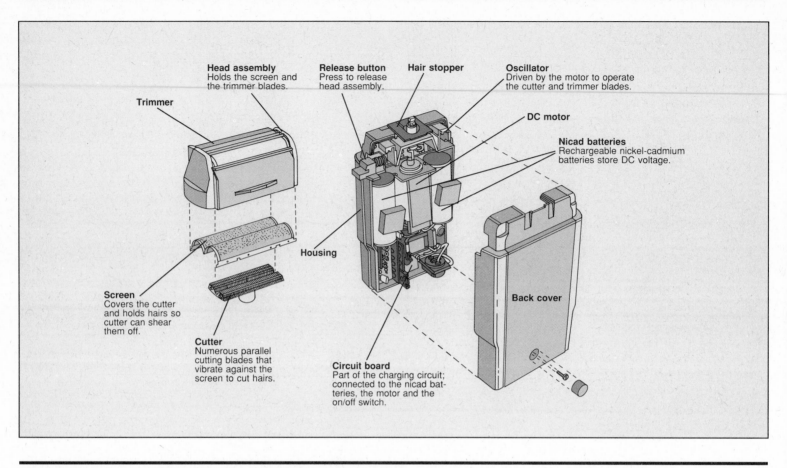

Head assembly Holds the screen and the trimmer blades.

Release button Press to release head assembly.

Hair stopper

Oscillator Driven by the motor to operate the cutter and trimmer blades.

Trimmer

DC motor

Nicad batteries Rechargeable nickel-cadmium batteries store DC voltage.

Housing

Screen Covers the cutter and holds hairs so cutter can shear them off.

Cutter Numerous parallel cutting blades that vibrate against the screen to cut hairs.

Circuit board Part of the charging circuit; connected to the nicad batteries, the motor and the on/off switch.

Back cover

SERVICING THE TRIMMER AND CUTTER ASSEMBLIES

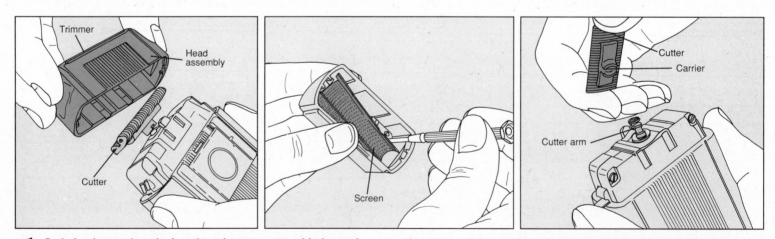

Trimmer

Head assembly

Cutter

Screen

Cutter

Carrier

Cutter arm

1 **Lubricating and replacing the trimmer, cutter blades and screen.** Turn off and unplug the shaver. If either the trimmer or the cutter doesn't work at all, remove the head assembly by depressing the release button and go to step 2 and service the oscillators. To lubricate a noisy trimmer, apply a couple of drops of manufacturer-recommended lubricant to the tiny blades and wipe off excess oil with a soft cloth. If this doesn't quiet the trimmer, or if its blades are broken or corroded, replace it. On the model shown here, the trimmer is part of the head assembly. Replace the entire assembly *(above, left)* with an exact replacement head assembly purchased from an authorized service center. If the shaver has been cutting poorly or pinching, inspect the mesh screen for tears and corrosion. To replace the screen, remove the head and pop

out the screen; if it catches, use the tip of a small screwdriver to pry its edges free *(above, center)*. Install an exact replacement screen purchased from an authorized service center. Gently push the screen into the head until it catches.

Inspect the cutter blades for bends, nicks and corrosion. Remove a damaged cutter by giving it a quarter turn right or left to pop it off the tip of the cutter arm *(above, right)*. Turn it over and use a screwdriver to pry off the carrier. Buy an identical cutter from an authorized service center and snap in the old carrier, centering it on the cutter. Install the cutter by sliding the carrier down onto the cutter arm as far as it will go. Then lock the cutter by rotating it a quarter turn. Snap on the head assembly.

SERVICING THE TRIMMER AND CUTTER ASSEMBLIES (continued)

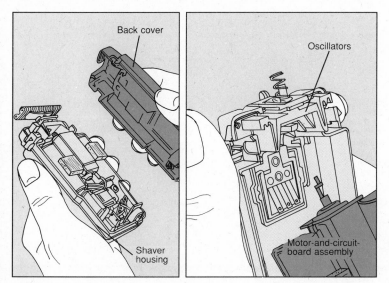

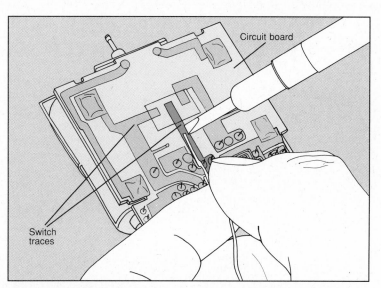

2 **Removing the back cover and accessing internal components.** Locate and remove all screws securing the back cover to the shaver housing. To find concealed screws, you may have to use the tip of a tiny screwdriver to pry out a plastic plug in the back of the shaver *(page 115)*. Lift off the back cover *(above, left)*, taking care not to pull any wires off the circuit board. Pop off the cutter by pushing it down and turning it one-quarter turn. To remove the motor-and-circuit-board assembly *(above, right)* first free the motor shaft from the hole in the bottom of the oscillators by lifting the assembly and twisting it sideways.

3 **Replacing the trimmer and cutter oscillators.** Remove the rubber hair stopper that fits over the oscillators. Slide the cutter oscillator, then the trimmer oscillator, out of the slots in the top edges of the shaver housing. Inspect them for wear, particularly around the shaft holes and edges *(above)*. If either oscillator is damaged, replace it and the cutter carrier *(page 62)* with duplicates purchased from an authorized service center. Install the oscillators and the cutter carrier, and reassemble the shaver by reversing the steps taken to disassemble it. Cold check for leaking voltage *(page 114)*.

SERVICING THE ON/OFF SWITCH (Rechargeable type)

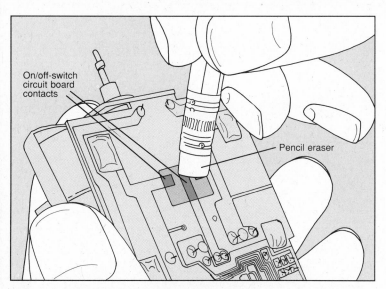

1 **Inspecting and cleaning switch contacts.** Some foil head shavers, such as the one shown here, have an on/off switch with three exposed contacts on the circuit board and three corresponding leaf contacts on the inside back cover. To inspect the switch, remove the back cover of the shaver and lift out the motor-and-circuit-board assembly *(step above)*. If any leaf contacts are bent or broken, replace the back cover with a duplicate purchased from an authorized service center. Clean dirty or tarnished contacts by rubbing them lightly with a pencil eraser *(above)*, then blowing away the rubbings.

2 **Repairing switch traces.** Examine the copper circuit board traces that connect the switch contacts to other circuit board components and repair any breaks in the traces. If there is a varnish-like or plasticized protective coating over a broken trace, gently use the tip of a utility knife to scrape away the coating, then use rosin-core solder and a soldering iron *(page 121)* to close the break in the trace *(above)*. After repairing switch traces, reassemble the shaver, reversing the steps taken to disassemble it. If the shaver still doesn't work, take it for professional service.

TESTING AND REPLACING THE MOTOR (Rechargeable type)

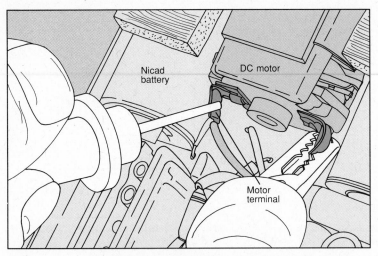

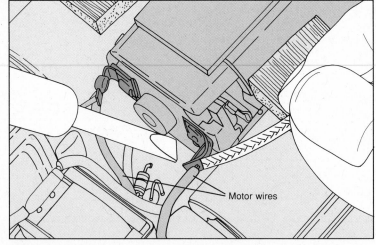

1 **Testing voltage to the motor.** Remove the head assembly, the back cover and the motor-and-circuit-board assembly *(page 63, top, step 2)*. Resolder any motor wires that are disconnected *(page 121)*. Plug the power cord into the shaver. Set a multitester to 10 volts DCV *(page 113)*. Clip a tester probe to one of the motor terminals. Plug in the shaver and turn it on. Without touching the shaver with your hands, touch the second probe to the second motor terminal *(above)*. If the tester needle dips to the left, reverse the probes. If the tester reads 2.5 to 3 volts, there is sufficient voltage going to the motor; suspect a faulty motor and replace it *(step 2)*. If the tester shows less than 2.5 volts, suspect faulty nicad batteries and test the charging circuit *(page 122)*.

2 **Replacing the motor.** Label the motor wires, then desolder them from their terminals *(page 120)*, using desoldering braid and a soldering iron *(above)*. The motor in the model shown here is held to the circuit board with adhesive padding and may simply be pulled off. Remove the padding from the motor. Buy an exact replacement motor from an authorized service center. Reinstall the padding, install the new motor on the circuit board and solder the wires to the motor terminals *(page 121)*. Remove the labels and reassemble the shaver, reversing the steps taken to disassemble it. Cold check for leaking voltage *(page 114)*.

ROTARY HEAD SHAVERS

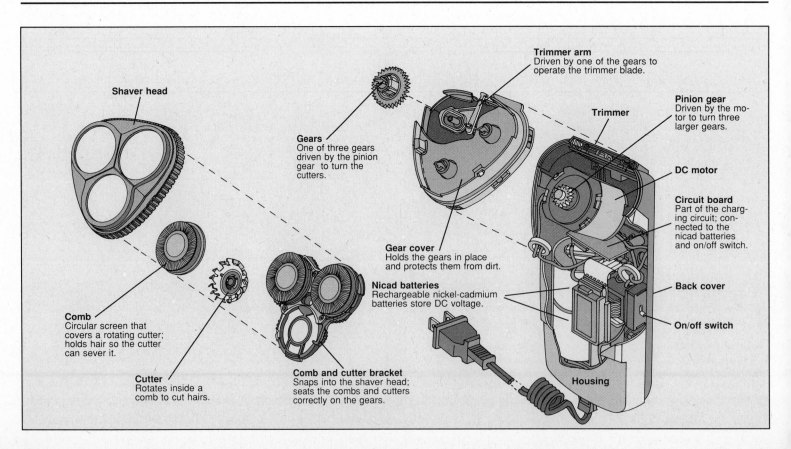

Shaver head

Trimmer arm
Driven by one of the gears to operate the trimmer blade.

Trimmer

Pinion gear
Driven by the motor to turn three larger gears.

Gears
One of three gears driven by the pinion gear to turn the cutters.

DC motor

Circuit board
Part of the charging circuit; connected to the nicad batteries and on/off switch.

Gear cover
Holds the gears in place and protects them from dirt.

Nicad batteries
Rechargeable nickel-cadmium batteries store DC voltage.

Back cover

On/off switch

Comb
Circular screen that covers a rotating cutter; holds hair so the cutter can sever it.

Cutter
Rotates inside a comb to cut hairs.

Comb and cutter bracket
Snaps into the shaver head; seats the combs and cutters correctly on the gears.

Housing

CLEANING AND REPLACING COMBS AND CUTTERS

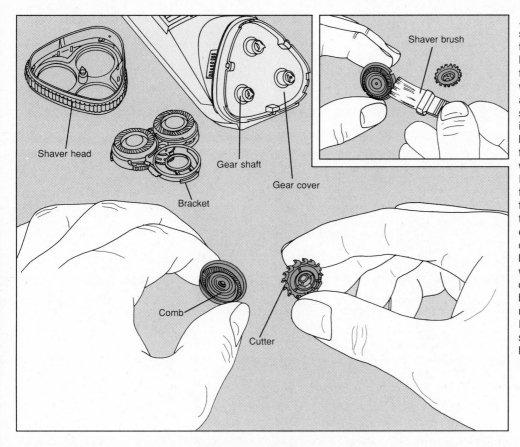

Access and cleaning. Turn off and unplug the shaver and remove its detachable cord. Hold the shaver over newspaper to catch loose hair and pull off the shaver head. Wipe clean the shaver head, gear cover and gear shafts with a soft cloth. Turn over the shaver head and unlock the comb-and-cutter bracket inside the top: On the model shown here, push in the small plastic dial in the center of the bracket and turn it counterclockwise. Lift out the bracket. If the combs and cutters come with it, slide them out of their retaining clips. If not, remove them separately. Take the cutters out of their combs *(left)*, but make sure to keep each pair together; interchanging them may damage them, producing an uneven shave. To clean the combs and cutters of hair and dirt, use a small stiff-bristled brush *(inset)*; many shaver kits are supplied with one. Examine the combs and cutters for chips or corrosion. If any comb or cutter is broken, replace all pairs. Buy exact replacements from an authorized service center. Install the combs and cutters, reversing the steps taken to remove them, and snap the head back onto the shaver housing.

ACCESSING AND SERVICING THE GEARS

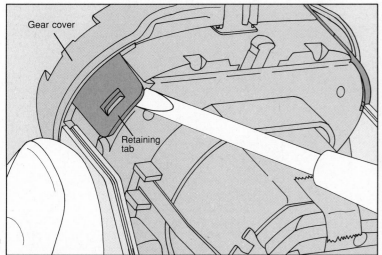

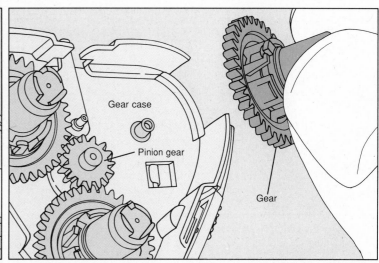

Accessing and servicing gears. Turn off and unplug the shaver and remove its detachable cord. Hold the shaver over newspaper to catch loose hair. Unscrew and remove the back cover and pull off the shaver head. To remove the gear cover, use a screwdriver to pry up the two plastic retaining tabs *(above, left)* and release the third tab by pulling up firmly on the front of the cover. Shake loose dirt and hair out of the gears. Lift the larger gears out of their seats *(above, right)*. Clean the gears and gear case with a soft cloth dipped in rubbing alcohol. Inspect the gears for worn, cracked or missing teeth. If the pinion gear is damaged, remove it by gently prying it off the end of the motor shaft with a tiny screwdriver. Replace a worn or broken gear or pinion gear with a duplicate purchased at an authorized service center. Snap the new pinion gear onto the end of the motor shaft. Before installing the larger gears, apply a drop of light machine oil on the metal seating pins on their undersides. Reseat the gears and check that they turn smoothly by rotating them with your fingers. Add a drop of machine oil to the top of the pinion gear. Snap on the gear cover, aligning its flat edge with the back of the shaver, and reassemble the shaver.

SERVICING THE TRIMMER ASSEMBLY

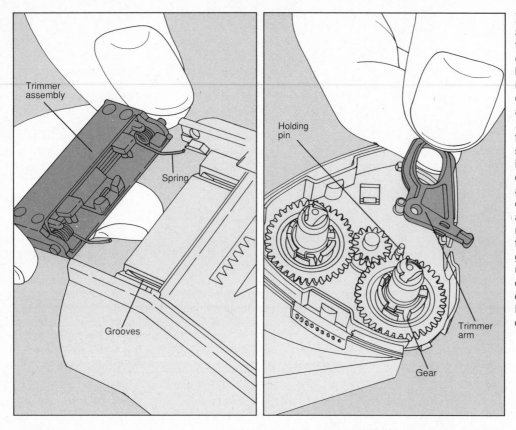

Inspecting and replacing the trimmer and trimmer arm. Turn off and unplug the shaver. Unscrew and remove the back cover. Snap open the trimmer and inspect its blades. If any are bent, broken or corroded, replace the trimmer. Gently use a tiny screwdriver to pry out one corner of the trimmer assembly, then pull it free of the cover *(far left)*. Buy an exact replacement from an authorized service center and snap it in place so that the ends of its springs slide smoothly into the grooves in the cover. If the trimmer doesn't work at all, suspect a worn trimmer arm. Access the gears by removing the gear cover *(page 65)*. Rotate the gears by hand and check whether the tip of the trimmer arm moves from side to side. If it doesn't, the trimmer arm is worn; replace it. Lift out the gear that sits on top of it, then pull the arm up off its holding pin *(near left)*. Buy a duplicate trimmer arm from an authorized shaver center. Install the new trimmer arm on its pin. Reinstall the gear, the gear cover, the back cover and the shaver head.

SERVICING THE ON/OFF SWITCH (Rechargeable type)

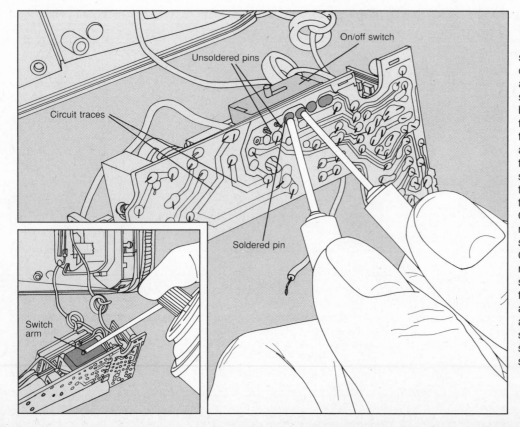

1 Testing and cleaning the on/off switch. Turn off the shaver, unplug it and remove the detachable power cord. Unscrew and remove its back cover. Lift out the circuit board, taking care not to disconnect any wires. Locate the on/off switch on one side of the circuit board. Turn over the circuit board and note the four soldered pins and the two unsoldered pins on the light-colored trails, or circuit traces, of the board; these are the switch terminals. To test a switch like the one on the model shown here, first desolder *(page 120)* one of the wires attached to the nicad batteries. Set a multitester to test continuity *(page 113)*. Turn the switch to OFF and touch a tester probe to each terminal pin of the middle pair *(left)*; the tester should show no continuity. Turn the switch to ON and touch the probes to the same terminals; the tester should show continuity. If the switch fails either test, spray electrical contact cleaner inside *(inset)*. Work the switch arm back and forth and then test the switch again. If the switch tests OK, reassemble the shaver, reversing the steps taken to disassemble it. If the switch still tests faulty, go to step 2 to replace it.

SERVICING THE ON/OFF SWITCH (Rechargeable type, continued)

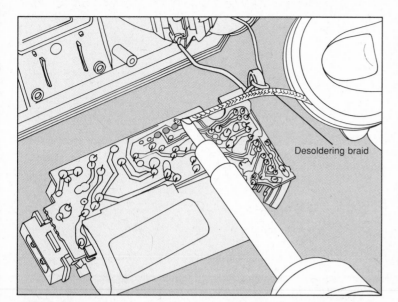

Desoldering braid

2 **Removing the on/off switch.** Sketch a diagram showing which terminal pins are soldered and which are not for correct reassembly. Use a soldering iron and desoldering braid *(above)* to desolder *(page 120)* the four soldered terminal pins. Use the braid to clean any solder residue off the ends of the pins and the circuit board. Pull the switch off the circuit board. Buy an exact replacement switch from an authorized service center and go to step 3 to install it.

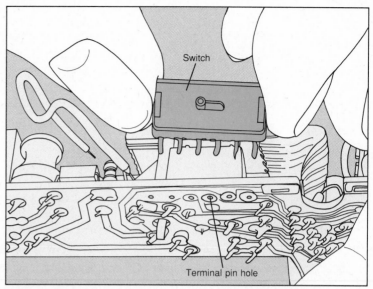

Switch

Terminal pin hole

3 **Installing an on/off switch.** Position the new switch over the terminal pin holes *(above)* and push it firmly into place on the circuit board. Consult the diagram you sketched earlier to determine which pins should be soldered. Using a soldering iron and rosin-core solder, solder *(page 121)* the tip of each of those terminal pins to the circuit board and resolder the disconnected wire to the nicad batteries. Fit the circuit board assembly back into the shaver housing and reassemble the shaver, reversing the steps taken to disassemble it. Cold check for leaking voltage *(page 114)*.

TESTING AND REPLACING THE MOTOR (Rechargeable type)

DC motor

Motor wires

1 **Testing voltage to the motor.** Unscrew the back cover and lift out the circuit board assembly, taking care not to disconnect any wires. Repair loose or broken wire connections *(page 121)*. Plug the power cord into the shaver. Set a multitester to 10 volts DCV *(page 113)*. Clip a tester probe to one of the motor terminals. Plug in the shaver and turn it on. Taking care not to touch the shaver with your hands, touch the second probe to the second motor terminal *(above)*. If the tester needle dips to the left, reverse the probes. If the tester reads 2.5 to 3 volts DC, there is sufficient voltage going to the motor; suspect a faulty motor and replace it *(step 2)*. If the meter reads less than 2.5 volts, suspect faulty nicad batteries and test the charging circuit *(page 122)*.

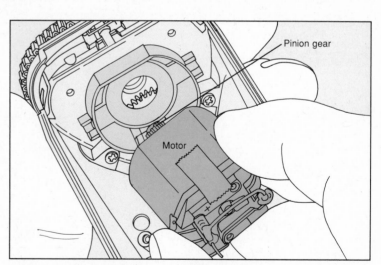

Pinion gear

Motor

2 **Replacing the motor.** Label the motor wires. Use desoldering braid and a soldering iron to desolder the wires from the motor terminals *(page 120)*. Disengage the motor from the retaining clips and pull it out *(above)*. Gently pry the pinion gear off the tip of the motor shaft with a small screwdriver. Buy an exact replacement motor from an authorized service center. Install the new motor by reversing these steps, then solder the wires from the circuit board to the motor terminals *(page 121)*. Reassemble the shaver by reversing the steps taken to disassemble it, and cold check for leaking voltage *(page 114)*.

CAN OPENERS

The electric can opener is a classic example of America's love affair with gadgets. Not only can it open cans, it may also pop open bottle caps, slice open plastic bags, twist open stubborn jar tops and, more commonly, sharpen knives and scissors. Most can openers work on power from a wall outlet. Some are rechargeable and, for the intrepid camper, there are models that can be plugged into a car cigarette lighter.

Not suprisingly, the can opener's movable cutting and sharpening mechanisms most often require maintenance. To diagnose problems, consult the Troubleshooting Guide at right. Before disassembling the appliance, review the tips on disassembly in Tools & Techniques *(page 115)*.

To clean or replace the cutter, take off the lever arm. This may mean removing one or two screws or, on the typical model illustrated below, working the lever arm free of the key-shaped hole in the front cover by lifting the arm and sliding it all the way to one side. The cutter blade or wheel is screwed to the lever arm. When installing a new cutter, place the thicker side next to the housing. To take off the back cover, remove its recessed screws. Sometimes separate covers protect the on/off switch contacts and grindstone. If replacing the unit's power cord, note the way the old cord was knotted and secured for correct reassembly.

TROUBLESHOOTING GUIDE

SYMPTOM	PROCEDURE
Can opener doesn't run at all	Reset breaker or replace fuse *(p. 112)* □○; have outlet serviced
	Test power cord *(p. 116)* ▣○▲
	Repair wire connections *(p. 118)* ▣○
	Take can opener for service
Can opener runs intermittently	Test power cord *(p. 116)* ▣○▲
	Repair wire connections *(p. 118)* ▣○
	Service on/off switch *(p. 69)* ▣○
Can drops off can opener	Shim feed gear *(p. 69)* □○
Can doesn't turn but feed gear turns when can isn't engaged	Clean, replace feed gear *(p. 69)* □○
Can doesn't turn and feed gear doesn't turn when can isn't engaged	Replace idler gear *(p. 69)* ▣○
	Clean motor *(p. 125)* ▣○
Can turns, but cutter skips or doesn't cut can at all	Remove lever arm, clean or replace cutter *(left)* □○
Knife sharpener doesn't sharpen	Replace grindstone *(p. 69)* ▣○

DEGREE OF DIFFICULTY: □ Easy ▣ Moderate ■ Complex
ESTIMATED TIME: ○ Less than 1 hour ◑ 1 to 3 hours
▲ Special tool required

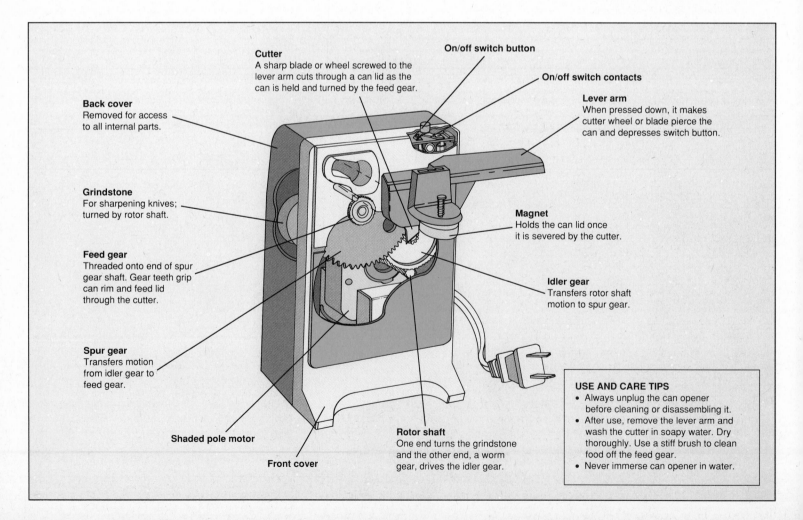

Cutter
A sharp blade or wheel screwed to the lever arm cuts through a can lid as the can is held and turned by the feed gear.

On/off switch button

On/off switch contacts

Back cover
Removed for access to all internal parts.

Lever arm
When pressed down, it makes cutter wheel or blade pierce the can and depresses switch button.

Grindstone
For sharpening knives; turned by rotor shaft.

Magnet
Holds the can lid once it is severed by the cutter.

Feed gear
Threaded onto end of spur gear shaft. Gear teeth grip can rim and feed lid through the cutter.

Idler gear
Transfers rotor shaft motion to spur gear.

Spur gear
Transfers motion from idler gear to feed gear.

Shaded pole motor

Rotor shaft
One end turns the grindstone and the other end, a worm gear, drives the idler gear.

Front cover

USE AND CARE TIPS
- Always unplug the can opener before cleaning or disassembling it.
- After use, remove the lever arm and wash the cutter in soapy water. Dry thoroughly. Use a stiff brush to clean food off the feed gear.
- Never immerse can opener in water.

HAND MIXERS

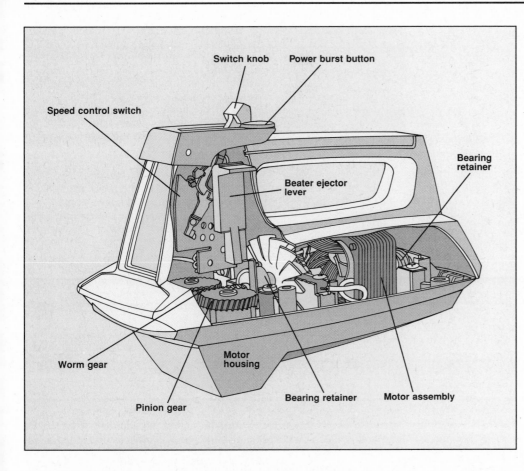

Accessing switches, motor and gears. Turn off and unplug the mixer. Remove the beaters and the detachable power cord, if it has one. Pry the switch knob off the lever. Turn over the mixer and remove the screws from the motor housing. Turn the mixer upright and work apart the housing, pulling the upper housing off over the beater ejector lever. You now have access to the motor and switch assembly. To access the gears, turn over the motor housing and pry off the small metal retaining rings and flat washers around the bottoms of the gear spindles, using a small screwdriver or long-nose pliers. Turn the motor housing upright. Unscrew the switch assembly from the top of the front bearing retainer and move the switch assembly to one side. You can now service the gears *(page 77).*

SERVICING THE SWITCHES (Hand mixers)

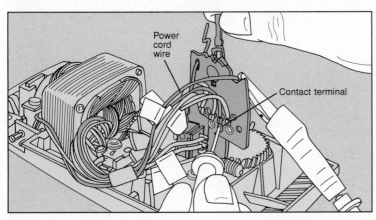

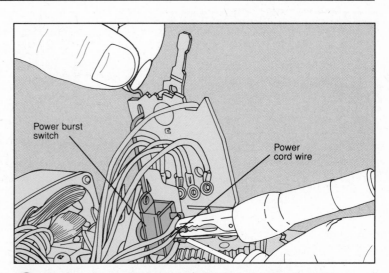

1 **Testing the speed control switch.** Turn off and unplug the mixer. Access the switches *(step above)*. Repair loose or broken wire connections *(page 118)*. Clean dirty switch contacts *(page 117)*. To test a speed control switch like the one shown here, use a continuity tester, or set a multitester to test continuity *(page 113)*. Identify the power cord wire connected to the switch and clip one tester probe to its terminal on the switch. Set the switch lever to the first contact terminal and touch the other probe to that terminal. The multitester should show continuity. Repeat the test at each switch lever setting, in turn *(above)*. If there is continuity at each setting, the switch is OK; go to step 2. If there is no continuity at a setting, replace the switch assembly *(step 3)*.

2 **Testing the power burst switch.** Use a continuity tester, or set a multitester to test continuity *(page 113)*. Identify the power cord wire connected to the power burst switch, located on the main switch assembly. Clip one tester probe to the wire's terminal on the switch. Then touch the second probe to one of the other two terminals as you depress the shaft of the power burst switch *(above)*. The multitester should show continuity. Repeat the test for the other terminal. If there is continuity, the switch is OK. If there is no continuity in a test, replace the switch assembly.

SERVICING THE SWITCHES (Hand mixers, continued)

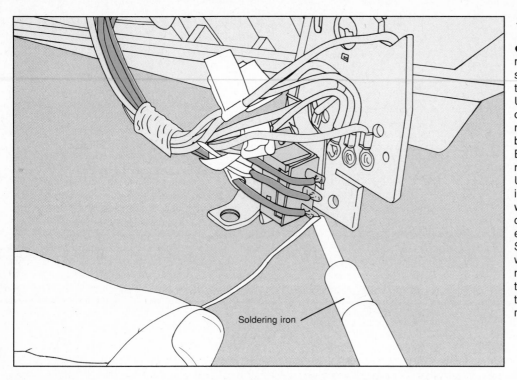

Soldering iron

3 **Replacing the main switch assembly.**
Label all main switch-assembly wires, noting their positions for correct reconnection, and disconnect them. On the model shown here, use a soldering iron *(page 120)* to desolder the power-burst switch wires. Use diagonal-cutting pliers to cut the speed-control switch wires as close to their terminals as possible. Unscrew the switch assembly from the top of the front bearing retainer. Buy an exact replacement from an authorized service center or the manufacturer. Use a wire stripper to strip a small piece of insulation off the tips of the speed control wires and slip each wire into its correct metal crimp connector on the new switch. Close each crimp connector with long-nose pliers. Solder *(page 120)* each power-burst switch wire to its correct terminal *(left)*. Secure the new switch assembly to the front bearing retainer. Install the upper housing, reversing the steps taken to remove it. Cold check the mixer for leaking voltage *(page 114)*.

GOVERNOR-TYPE STAND MIXERS

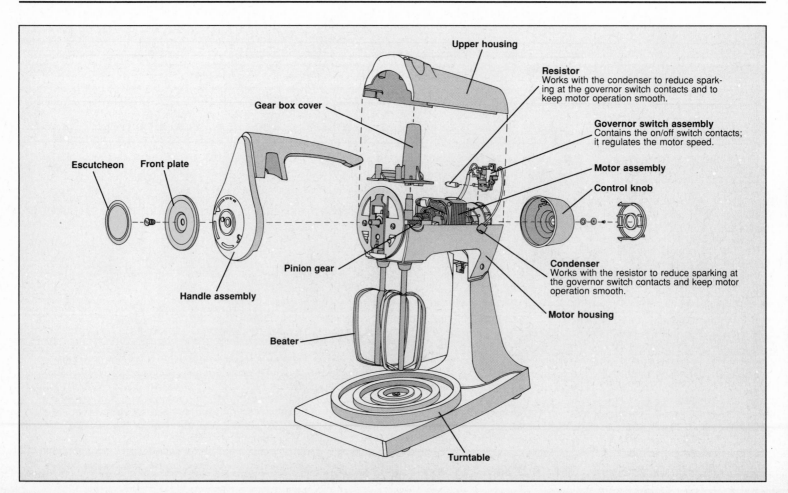

Upper housing

Resistor
Works with the condenser to reduce sparking at the governor switch contacts and to keep motor operation smooth.

Gear box cover

Governor switch assembly
Contains the on/off switch contacts; it regulates the motor speed.

Escutcheon

Front plate

Motor assembly

Control knob

Pinion gear

Handle assembly

Condenser
Works with the resistor to reduce sparking at the governor switch contacts and keep motor operation smooth.

Motor housing

Beater

Turntable

ACCESS TO INTERNAL PARTS (Governor-type stand mixers)

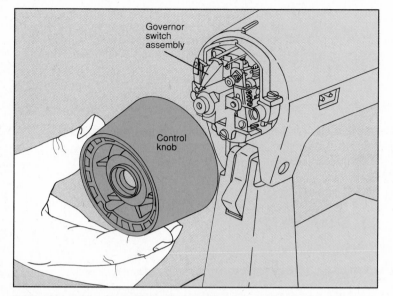

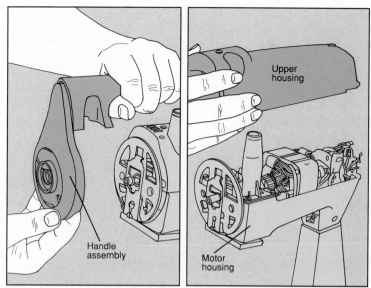

Removing and reinstalling the control knob. Turn off and unplug the mixer and remove the beaters. To remove the control knob on the model shown above, turn it to its highest setting, then use a screwdriver to pry the cap off its end. Note the order of the washers held by a screw to the center of the control knob. Remove the screw and catch the washers if they fall. Pull off the control knob (*above*). Reverse this sequence to reinstall the knob, taking care that the washers are put back in the correct order.

Removing and reinstalling the handle and upper housing. Remove the control knob (*left*). Then pry the escutcheon off the front of the mixer with the tip of a screwdriver. Unscrew the front plate and pull out the spring and washer beneath it. Grip the handle and rotate it counterclockwise as far it goes, then pull to release the handle assembly from the mixer (*above, left*). Unscrew the upper housing and lift it off the motor housing (*above, right*). Reverse this sequence to reinstall the housing and handle.

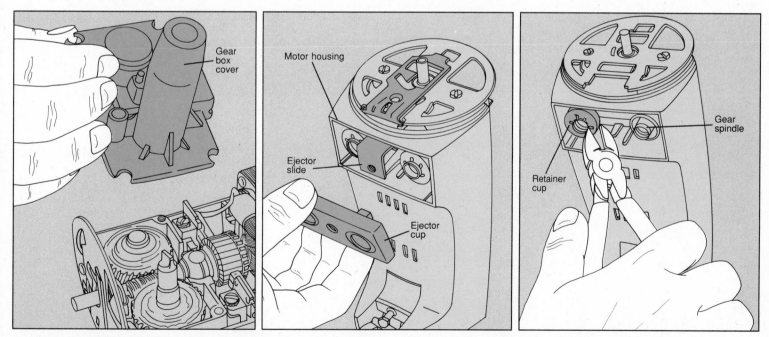

Accessing the gears. Remove the control knob (*step above, left*) and take off the handle and upper housing (*step above, right*). Unscrew and lift off the gear box cover (*above, left*). Tilt back the motor housing and unscrew and remove the ejector cup underneath it (*above, center*), then pull out the ejector slide. Use diagonal-cutting pliers to cut the retainer cups off the bases of the gear spindles

(*above, right*). Then remove the two spindle seals beneath the retainer cups. You can now service the gears (*page 77*). Before reassembling the mixer, purchase two identical retainer cups from the mixer manufacturer or an authorized service center to replace those you cut off. To put the mixer back together, reverse the sequence of steps taken to access the gears.

SERVICING THE GOVERNOR SWITCH ASSEMBLY AND CONDENSER (Governor-type stand mixers)

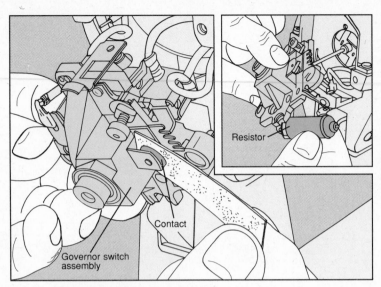

Resistor

Contact

Governor switch
assembly

1 **Servicing the governor switch assembly.** Remove the control knob *(page 73)*. Unscrew the governor switch from the back of the motor housing and pull it out of the mixer. Inspect the switch contacts. If the contacts can't close or are fused together, go to step 3 to replace the governor switch assembly. If the contacts are pitted or burned, clean them: Slip a small, folded piece of fine emery paper between them and gently rub their surfaces *(left)*. To remove the dust, pull a piece of plain paper between the contacts. After cleaning the contacts, inspect the resistor. If the resistor looks burned, flex one end of its bracket outward and pull it free *(inset)*. Replace it with an identical resistor purchased from an authorized service center.

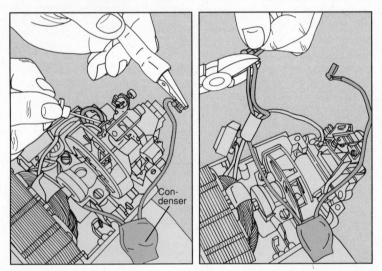

Con-
denser

2 **Testing and replacing the condenser.** Remove the handle and upper housing *(page 73)*. Use a continuity tester, or set a multitester to test continuity *(page 113)*. Locate the condenser and disconnect one of its wires. Touch a tester probe to each wire connector *(far left)*. The multitester should show resistance (although the needle may jump before it settles). If the condenser has continuity, replace it. Pull off the second wire connector and remove the condenser from the mixer. On the model shown here, a power cord wire is also attached to this wire connector. Use diagonal-cutting pliers to cut off the connector *(near left)*. Purchase an identical replacement condenser from an authorized service center or the mixer manufacturer. Install new wire connectors on the condenser wires *(page 118)*, making sure to reconnect the power cord wire if it was disconnected earlier. Reconnect the condenser to the governor switch. Reinstall the upper housing, handle and control knob *(page 73)* and cold check the mixer for leaking voltage *(page 114)*.

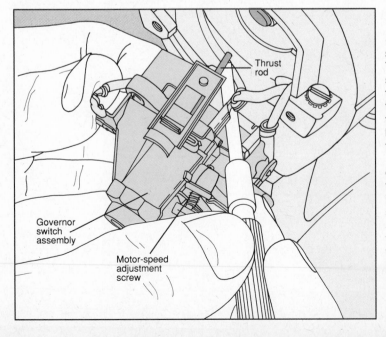

Thrust
rod

Governor
switch
assembly

Motor-speed
adjustment
screw

3 **Replacing the governor switch assembly.** Label the wires and pull their connectors off the switch terminals with long-nose pliers. Buy an identical replacement switch from an authorized service center or the manufacturer. To install the new switch, first reconnect the wires. Then insert the flat end of the thrust rod into the slot in the back of the new switch. Use a magnetized screwdriver to guide the free end of the thrust rod into the armature hole as you push the switch assembly back into place *(left)*. Screw the assembly onto the motor housing and reinstall the control knob. After reassembling the mixer, cold check for leaking voltage *(page 114)*. Then plug in the mixer and slowly rotate the control knob. If the mixer turns on at a higher speed setting than normal, or if the mixer doesn't turn off at all or reaches its maximum speed below the highest speed setting, the switch contacts need adjusting. Turn off the mixer, unplug it and remove the control knob. Use a small hex wrench to turn the motor-speed adjustment screw one-quarter turn: clockwise if the mixer turns on too late, and counterclockwise if it reaches maximum speed too soon. Reinstall the control knob and operate the mixer again. Repeat the adjustment until the mixer turns on at the correct setting, turning the adjustment screw no more than one-quarter turn each time.

ELECTRONIC STAND MIXERS

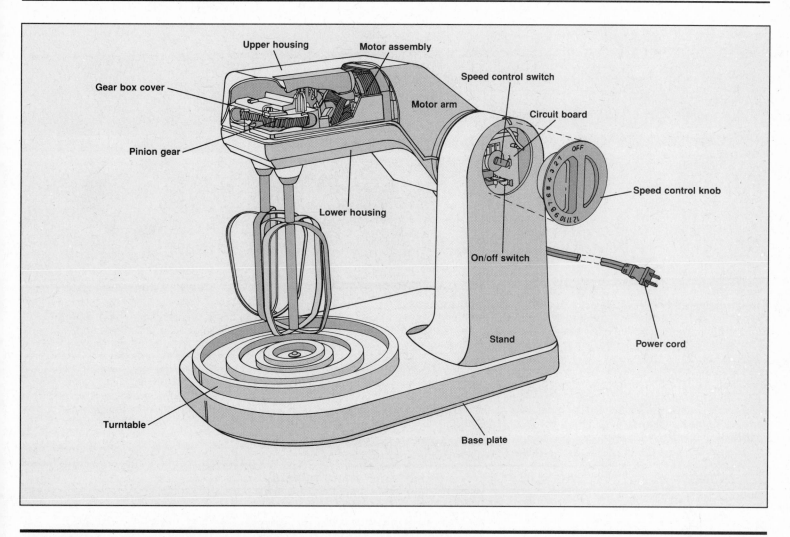

Upper housing

Motor assembly

Gear box cover

Speed control switch

Motor arm

Circuit board

Pinion gear

Speed control knob

Lower housing

On/off switch

Stand

Power cord

Turntable

Base plate

ACCESSING INTERNAL COMPONENTS (Electronic stand mixers)

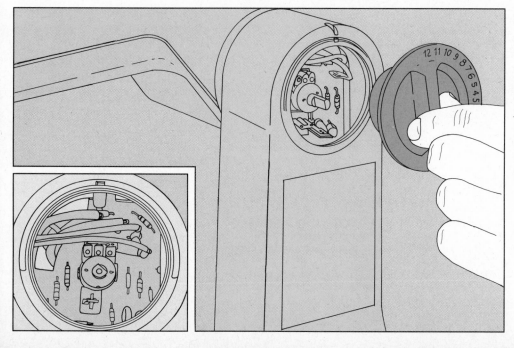

Removing and reinstalling the speed control knob. Turn off and unplug the mixer and remove the beaters. Turn the speed control knob all the way to its highest setting. Using a screwdriver, carefully pry the knob away from the mixer stand, then pull off the knob by hand *(left)*. You can now clean the on/off switch contacts and test the fuse-and-motor circuit *(page 76)*. When reinstalling the speed control knob, make sure that the circuit board is in place *(inset)* and turn the speed control knob to its highest setting before sliding it back on its shaft.

ACCESSING INTERNAL COMPONENTS (Electronic stand mixers, continued)

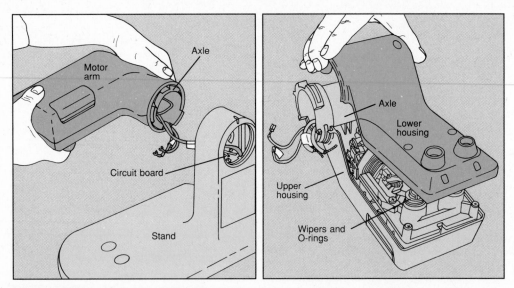

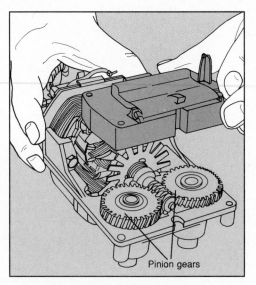

Disassembling and reassembling the motor arm. Turn off and unplug the mixer. Remove the control knob *(page 75)*. Label the wires and pull them off the circuit board. Reach through the speed-control knob hole and slide the circuit board down into the lower part of the stand until you can see the bracket that holds the motor arm axle to the stand. Unscrew the bracket and lift it out. Lift the motor arm up and back until the axle snaps free, then pull the arm off the stand *(above, left)*. Turn the motor arm upside down. Unscrew and lift off the lower housing *(above, right)*, and remove the motor-and-gear assembly from the upper housing. Keep track of the ejector spring that may fall off the bottom of the motor-and-gear assembly. Before reassembling the motor arm, roll the axle toward the back of the housing as far as it will go. Then reassemble the mixer, reversing the sequence of steps taken to disassemble it, and cold check for leaking voltage *(page 114)*.

Accessing the gears. Turn off and unplug the mixer. Disassemble the motor arm *(left)*. Remove the felt wipers and O-rings from the bottoms of the gear spindles and put them aside for reassembly. Remove the screws holding the gear box cover to the bottom of the motor-and-gear assembly. Then turn over the assembly and lift off the gear box cover *(above)*. After completing repair, reverse the sequence of steps taken to access the gears.

SERVICING THE ELECTRICAL COMPONENTS (Electronic stand mixers)

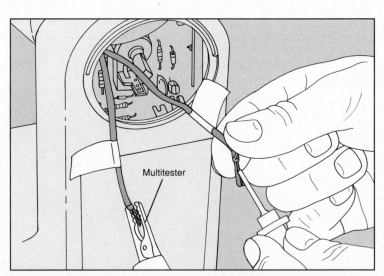

Servicing the on/off switch. Remove the speed control knob *(page 75)*. Inspect the contacts. If they are not touching, use long-nose pliers to bend them gently until they do. If the contacts are pitted, burned or dirty, clean them by slipping a small, folded piece of fine emery paper between the upper and lower contacts. Gently rub their surfaces *(above)*, then rub them with a piece of plain paper. After servicing the switch, reinstall the speed control knob and cold check the mixer for leaking voltage *(page 114)*. If the contacts can't be repaired or cleaned, replace the circuit board *(page 77)*.

Testing the fuse-and-motor circuit. Turn off and unplug the mixer. Remove the speed control knob *(page 75)*. Locate the fuse lead and motor lead on the circuit board; they are distinguished from the power cord leads by their smaller gauge. Label the leads and pull them off the circuit board. Set a multitester to RX10 *(page 113)* and touch a tester probe to the end of each lead *(above)*. The multitester should show very low resistance. If it doesn't, suspect either a faulty fuse or a faulty motor. Test the fuse *(page 77)* to find out which one needs repair. If the fuse and motor test OK, repair any loose or broken wire connections *(page 118)*.

SERVICING THE ELECTRICAL COMPONENTS (Electronic stand mixers, continued)

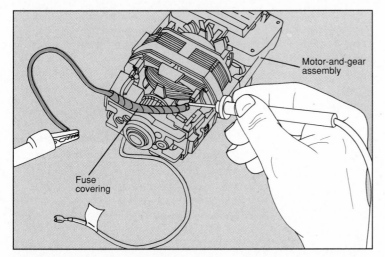

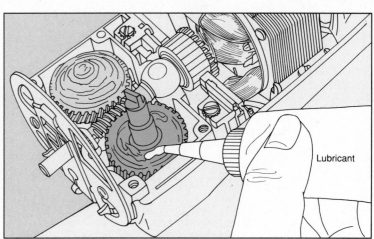

Testing and replacing the fuse. Turn off and unplug the mixer. Remove the speed control knob *(page 75)* and disassemble the motor arm *(page 76)*. Turn the motor-and-gear assembly over on its top. Locate the fuse; it may be hidden by a protective covering. Use a continuity tester, or set a multitester to test continuity *(page 113)*. Touch a probe to the wire connector on the end of each fuse lead. The multitester should show continuity. If the fuse tests OK, suspect a faulty motor and service it *(page 123)*. If the fuse tests faulty, remove it by pulling its lead off the motor terminal. Buy an exact replacement fuse from the mixer manufacturer or from an authorized service center. Connect the fuse to the motor and reassemble the mixer by reversing the steps taken to access the fuse. Cold check the mixer for leaking voltage *(page 114)*.

Replacing the circuit board. Turn off and unplug the mixer. Remove the speed control knob *(page 75)*. Label all wires to the circuit board, and disconnect them. Reach through the speed-control knob hole and slide the circuit board down into the lower part of the stand as far as you can. Lay the mixer on its side. Unscrew and remove the base plate; on the model shown here, peel off the rubber foot pads to access the base plate screws underneath. Remove the screws from the channel support and lift it off the bottom of the stand. Slide out the circuit board *(above)*. Buy a replacement circuit board from the mixer manufacturer or an authorized service center. Install the new board by reversing the sequence of steps taken to remove the old one. Cold check the mixer for leaking voltage *(page 114)*.

SERVICING THE GEARS (All mixers)

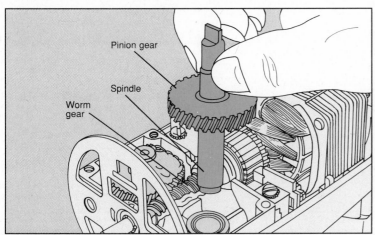

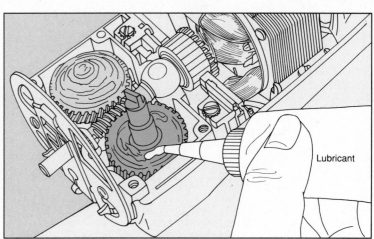

Cleaning and replacing the gears. Access the mixer gears *(hand mixers, page 71; governor-type stand mixers, page 73; electronic stand mixers, page 76)*. Pull out the pinion gears *(above)* and inspect them for wear or cracks. Replace both gears if either is damaged. Buy identical replacement gears and the recommended gear lubricant from an authorized service center. Clean the gear box thoroughly with a soft cloth, and use a small brush to remove plastic shavings and old lubricant from the grooves of the worm gear. Seat the new gears in place. Then, while you hold them firmly in position with one hand, tip back the mixer housing and snap the beaters into the gear spindles underneath. If the beater blades intersect one another at a 45-degree angle, the gears are seated

properly. If not, remove one of the beaters and lift out, rotate and reseat its gear. Snap the beater back in and check the beater positions again. Repeat until the beater blades intersect at a 45-degree angle. Once they do, secure the gears in place by reinstalling any washers, seals and clips. If the old gears were lubricated and housed in a separate box, apply a generous amount of the recommended lubricant around the teeth of the gears and on their tops *(above)*, or use a multipurpose, high-temperature, high-pressure grease. Replace the gear box cover, if it has one, and reassemble the mixer, reversing the sequence of steps taken to access the gears. Cold check the mixer for leaking voltage *(page 114)*.

BLENDERS

The blades of a blender are driven by a small universal motor directly below the jar and blade assembly. A push-button switch regulates a variety of speeds, ranging from chopping to liquefying.

The illustration below shows a typical blender with a removable blade assembly. The jar base, blades and seal ring can be taken apart for cleaning. When washing the blender, inspect the jar and blade assembly for cracks, bent blades or a dried-out seal ring; replace damaged parts. Wash the blade assembly immediately after each use to keep dried food from jamming its shaft. Inspect the blade socket that fits onto the drive stud, and replace the blade if it is stripped. Once a month, lubricate its moving metal parts with mineral oil; do not apply machine oil to parts that come in contact with food.

Blender problems are typically caused by overloading the jar. This slows the blades and puts a strain on the motor, causing it to overheat and burn out. Turn off the blender immediately if it begins to hum or has a burning odor—replacing the motor can cost as much as replacing the blender.

Testing a multi-speed switch is complicated. To determine whether the switch is at fault, consult the Troubleshooting Guide and rule out all other possible causes first. To service a simple on/off switch, see Tools & Techniques *(page 117)*. After repair, cold check for leaking voltage *(page 114)*.

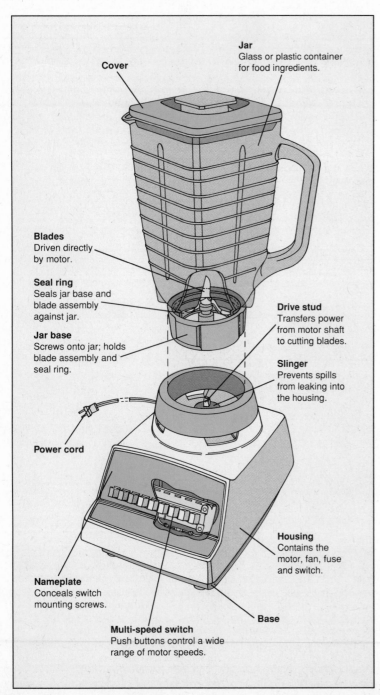

Cover

Jar
Glass or plastic container for food ingredients.

Blades
Driven directly by motor.

Seal ring
Seals jar base and blade assembly against jar.

Jar base
Screws onto jar; holds blade assembly and seal ring.

Drive stud
Transfers power from motor shaft to cutting blades.

Slinger
Prevents spills from leaking into the housing.

Power cord

Housing
Contains the motor, fan, fuse and switch.

Nameplate
Conceals switch mounting screws.

Base

Multi-speed switch
Push buttons control a wide range of motor speeds.

TROUBLESHOOTING GUIDE

SYMPTOM	PROCEDURE
Blender doesn't run at all	Reset breaker or replace fuse *(p. 112)* □○; have outlet serviced
	Test power cord *(p. 116)* ◼○▲
	Clean or replace switch *(p. 79)* □◓
	Remove base *(p. 79)* □○; repair faulty wire connections *(p. 118)* ◼○
	Test and replace fuse *(p. 79)* □○▲
	Test motor *(p. 80)* ◼○▲
Blender runs intermittently	Test power cord *(p. 116)* ◼○▲
	Remove base *(p. 79)* □○; repair faulty wire connections *(p. 118)* ◼○
	Replace brushes *(p. 124)* ◼◓
Blender doesn't run at some speeds	Remove base *(p. 79)* □○; repair wire connections *(p. 118)* ◼○
	Test motor *(p. 80)* ◼○▲
	Clean or replace switch *(p. 79)* □◓
Motor hums but blades don't turn	Reduce load in jar
	Wash blade assembly and lubricate with mineral oil; or replace
	Inspect and replace blade assembly; replace drive stud *(p. 80)* □◓
	Test motor *(p. 80)* ◼○▲
Blender vibrates noisily	Inspect and replace blade assembly; replace drive stud *(p. 80)* □◓
	Tighten drive stud or fan locknut *(p. 80)* □○
	Test motor *(p. 80)* ◼○▲
Jar leaks	Tighten base onto jar
	Inspect and replace jar, seal ring or blade assembly
Blender overheats	Reduce load in jar
	Clean air screen in base
	Test motor *(p. 80)* ◼◓▲
Motor fuse blows repeatedly	Test motor *(p. 80)* ◼◓▲

DEGREE OF DIFFICULTY: □ Easy ◼ Moderate ◼ Complex
ESTIMATED TIME: ○ Less than 1 hour
◓ 1 to 3 hours ● Over 3 hours
▲ Special tool required

SERVICING THE MULTI-SPEED SWITCH

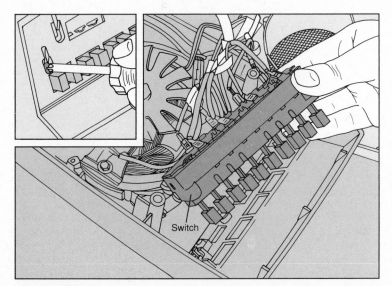

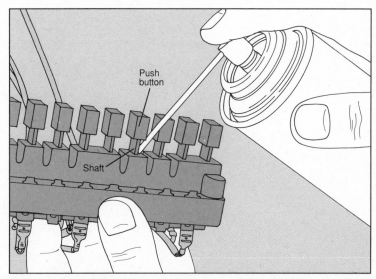

1 **Removing the base and accessing the switch.** Unplug the blender and remove the jar. Turn over the blender and remove the screws securing the base to the housing. Pull off the base to expose the switch, fan, motor and fuse. To service the switch, turn the blender upright and pry off the nameplate with a utility knife. Remove the mounting screws concealed beneath it *(inset)*, turn over the blender again and lift out the switch *(above)*.

2 **Cleaning or replacing the switch.** Scrape off hardened food deposits and wipe the push buttons with a damp cloth. Spray electrical contact cleaner into the button openings *(above)*, pressing each button two or three times. Clean the button shafts with a toothbrush and contact cleaner. If the buttons are jammed or the switch is faulty, label and disconnect the wires. Buy an exact replacement from an authorized service center and install it, making sure all wires are properly reconnected. Reassemble the blender and cold check for leaking voltage *(page 114)*.

TESTING AND REPLACING THE FUSE

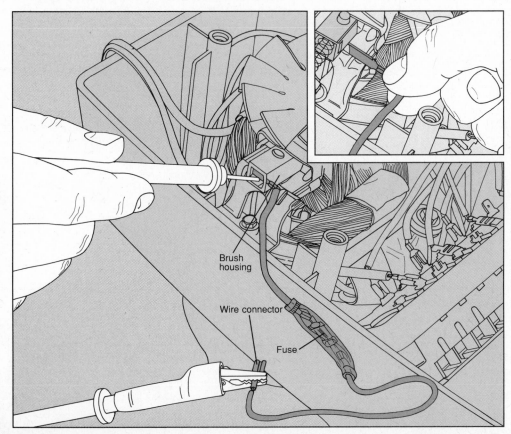

Testing and replacing the fuse. Unplug the blender and remove the jar. To access the fuse, remove the base *(step 1, above)*. Locate the fuse—usually encapsulated in clear plastic—and disconnect its wire from the switch. Set a multitester to RX1 *(page 113)* and clip a probe to the wire connector. Touch the other probe to the fuse terminal at the brush housing *(left)*. The tester should show continuity. If it doesn't, replace the fuse. Use long-nose pliers to straighten the fuse terminal and pull it out of the brush housing *(inset)*. The brush spring will be released; immediately place your finger over the opening and remove the terminal slowly, so the spring and its motor brush stay in the housing.

Buy an exact replacement fuse from the manufacturer or an authorized service center. Use a small screwdriver to push the spring into the brush housing, then insert the the new fuse's terminal into the housing slot. Bend back the end of the terminal to secure it. Connect the other wire end to the switch. Reassemble the blender and cold check for leaking voltage *(page 114)*.

SERVICING THE MOTOR

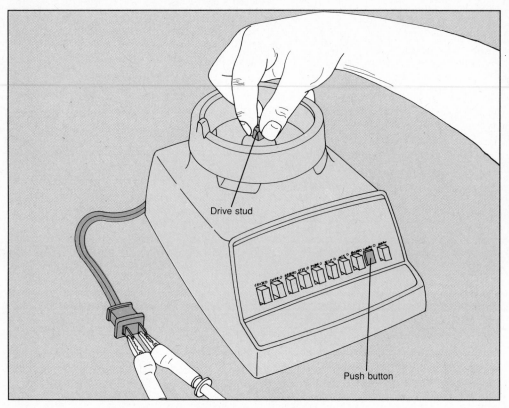

Drive stud

Push button

Testing the motor. Unplug the blender and remove the jar. Set a multitester at RX1 and clip a probe to each prong of the power cord plug. Press down one push button and read its resistance on the meter. With the button still depressed, rotate the drive stud one full turn *(left)*. If this causes the tester to register a change in resistance, service the motor *(page 123)*. Continue the test with each push button. Also note any deviation of more than about 15 ohms between buttons; this indicates a problem with the motor windings or the switch itself.

Remove the base *(page 79, step 1)* and inspect the motor assembly. If the motor smells burned or the windings are charred, take the unit for professional service. Otherwise, replace the switch *(page 79, step 2)*.

SERVICING THE DRIVE STUD AND SLINGER

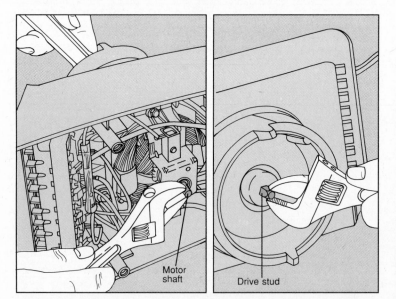

Motor shaft

Drive stud

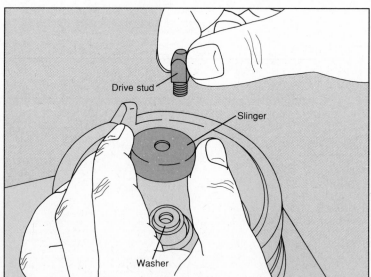

Drive stud

Slinger

Washer

1 **Freeing the drive stud.** Unplug the blender and remove the jar. Inspect the drive stud for worn edges; replace it if damaged. Remove the base *(page 79, step 1)* and lay the blender on its side. Locate the motor shaft. On the model shown, you must first remove the fan: Hold the fan with a rag, or wear a glove, and use an adjustable wrench to remove the fan locknut. Pull the fan and its washers off the shaft, taking care to lay the parts in correct order for reassembly. To free the drive stud, steady the motor shaft with the wrench *(above, left)*, while you use another wrench to unscrew the stud from the other side *(above, right)*.

2 **Replacing the drive stud.** Remove the drive stud and slinger, leaving the washer in place *(above)*. Use a toothbrush to clean the slinger of any hardened food deposits; replace it if rusted or worn. Buy exact replacement parts from the manufacturer or an authorized service center. Center the slinger and washer over the motor shaft, and insert the drive stud through them into the hole in the motor shaft. Tighten the stud with the two-handed technique you used to remove it, then reassemble the blender, reversing the steps you took to disassemble it. Cold check for leaking voltage *(page 114)*.

FOOD PROCESSORS

Food processors do what mixers and blenders do, only faster. Their powerful motors and razor-sharp blades can slice, shred or pulverize almost any food in a matter of seconds.

A belt-drive food processor *(page 82)* has a bowl capacity large enough to hold a shredded head of cabbage. The motor sits to one side of the bowl; a drive belt on a wheel-and-pulley assembly turns the blade. Switches control the speed and duration of the blade action. The direct-drive food processor is usually more compact. It houses the motor-and-gear assembly beneath the bowl; the blade is attached directly to the motor shaft or, as in the variation shown here *(page 84)*, the motor shaft drives a gear that turns the blade. Both types have a safety switch that activates the motor only when the cover is securely in position.

Most food processor malfunctions are caused by improper cleaning. Hardened food deposits in the drive shaft or spindle shaft cause noisy vibration, poor functioning and, eventually,

permanent damage to the blade and the gear or drive assembly. Leaks and spills can short the switches and motor. The Troubleshooting Guide below will help you locate the most likely cause of your food processor's problem and direct you to its repair. Motor problems are often covered by the manufacturer's warranty. If not, consult Tools & Techniques *(page 123)* to service the motor.

Take safety precautions when working on a food processor. Always unplug the power cord before beginning a repair, and handle the sharp cutting discs and blades with care. When disassembling an older model, check carefully inside the housing for a capacitor—a battery-like component usually mounted beside the motor. The capacitor stores a potentially dangerous electrical charge; discharge it *(page 114)* before attempting a repair. If your food processor is much different from the models shown in this chapter, or if it is still under warranty, have it serviced professionally.

TROUBLESHOOTING GUIDE

SYMPTOM	POSSIBLE CAUSE	PROCEDURE
Food processor doesn't work at all	No power to outlet or outlet faulty	Reset breaker or replace fuse *(p. 112)* □○; have outlet serviced
	Bowl and cover positioned incorrectly	Consult owner's manual for correct positioning
	Cam in edge of cover worn or broken	Inspect cover and replace it if damaged
	Power cord faulty	Test and replace power cord *(p. 116)* ▣○
	Fuse blown (direct-drive type)	Test and replace food processor fuse *(p. 85)* ▣○
	Motor faulty	Service motor *(p. 123)* ▣◗▲
	Wire connections loose or faulty	Tighten or repair wire connections *(p. 118)* ▣○
	Switches faulty	Test and replace multi-control switches *(belt-drive type, p. 83)* ▣◗; test and replace safety switch *(belt-drive type, p. 84)* ▣◗; or take food processor for professional service *(direct-drive type)*
Food processor runs on only one speed setting (belt-drive type)	Speed control switch faulty	Test and replace speed control switch *(p. 83)* ▣○
	Wire connections loose or faulty	Tighten or repair wire connections *(p. 118)* ▣○
	Motor faulty	Service motor *(p. 123)* ▣◗▲
Motor runs but blades don't turn	Drive belt broken (belt-drive type)	Replace drive belt *(p. 82)* □○
	Drive belt tension incorrect (belt-drive type)	Adjust drive belt *(p. 82)* □○
	Spindle shaft broken (belt-drive type)	Replace spindle shaft *(p. 82)* □○
	Gear stripped or broken (direct-drive type)	Replace gear *(p. 85)* ▣◗
Blades slow down and speed up erratically (belt-drive type)	Drive belt worn	Replace drive belt *(p. 82)* □○
	Drive belt tension incorrect	Adjust drive belt *(p. 82)* □○
Food processor is noisy or vibrates excessively	Blades or blade shaft dirty, worn or broken	Clean or replace blade assembly
	Spindle shaft dirty or worn (belt-drive type)	Clean or replace spindle shaft *(p. 82)* □○
	Drive shaft dirty or worn (direct-drive type)	Clean or replace drive shaft *(p. 85)* □○
Food processor overheats	Bowl overloaded	Consult owner's manual for correct loading instructions
	Air intake screen clogged	Clean screen in lower housing with a toothbrush and vacuum cleaner
	Motor faulty	Service motor *(p. 123)* ▣◗▲
Food processor blows fuse or circuit breaker repeatedly	Household electrical circuit overloaded	Reduce number of appliances on circuit

DEGREE OF DIFFICULTY: □ Easy ▣ Moderate ■ Complex
ESTIMATED TIME: ○ Less than 1 hour ◗ 1 to 3 hours ● Over 3 hours

▲ Special tool required

BELT-DRIVE FOOD PROCESSORS

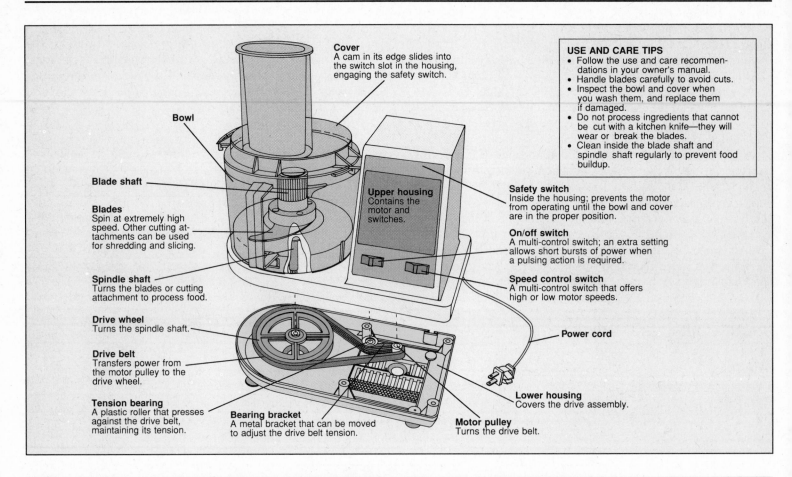

Cover
A cam in its edge slides into the switch slot in the housing, engaging the safety switch.

Bowl

Blade shaft

Blades
Spin at extremely high speed. Other cutting attachments can be used for shredding and slicing.

Spindle shaft
Turns the blades or cutting attachment to process food.

Drive wheel
Turns the spindle shaft.

Drive belt
Transfers power from the motor pulley to the drive wheel.

Tension bearing
A plastic roller that presses against the drive belt, maintaining its tension.

Bearing bracket
A metal bracket that can be moved to adjust the drive belt tension.

Upper housing
Contains the motor and switches.

Safety switch
Inside the housing; prevents the motor from operating until the bowl and cover are in the proper position.

On/off switch
A multi-control switch; an extra setting allows short bursts of power when a pulsing action is required.

Speed control switch
A multi-control switch that offers high or low motor speeds.

Power cord

Lower housing
Covers the drive assembly.

Motor pulley
Turns the drive belt.

USE AND CARE TIPS
- Follow the use and care recommendations in your owner's manual.
- Handle blades carefully to avoid cuts.
- Inspect the bowl and cover when you wash them, and replace them if damaged.
- Do not process ingredients that cannot be cut with a kitchen knife—they will wear or break the blades.
- Clean inside the blade shaft and spindle shaft regularly to prevent food buildup.

SERVICING THE DRIVE ASSEMBLY (Belt-drive food processors)

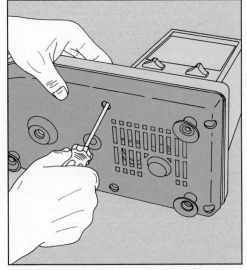

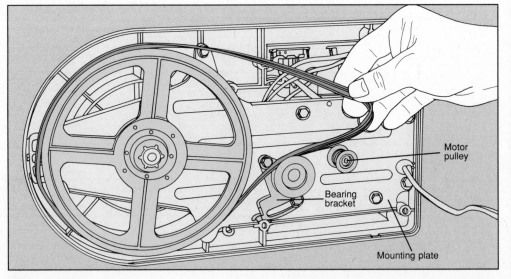

Motor pulley

Bearing bracket

Mounting plate

1 **Removing the lower housing.** Turn off and unplug the food processor. Take off the bowl and cover and lay the unit on its side. Remove the screws securing the lower housing *(above)* and lift it off to expose the drive assembly *(step 2)*. If you are simply adjusting the drive belt, go directly to step 4.

2 **Removing the drive belt.** Loosen, but do not remove, the screw securing the bearing bracket to the mounting plate. Shift the bracket to slacken the belt, then slip the belt up off the motor pulley and drive wheel *(above)*. Inspect the belt carefully; if it is cracked, worn, frayed or stretched, replace it. Purchase an identical replacement belt from the manufacturer or an authorized service center. If the blade assembly vibrates noisily when operating, go to step 3 to service the spindle shaft. Reinstall the belt by reversing the steps you took to remove it, and adjust its tension *(step 4)*.

SERVICING THE DRIVE ASSEMBLY (Belt-drive food processors, continued)

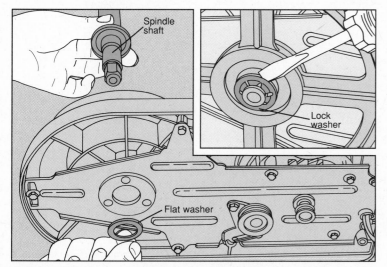

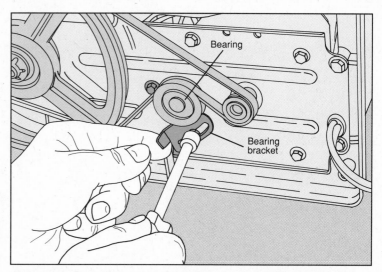

3 **Cleaning and replacing the spindle shaft.** Using a screwdriver, pry off the lock washer holding the spindle shaft to the center of the drive wheel *(inset)*. Remove the drive wheel and the flat washer beneath it, then pull the spindle shaft out through the top of the food processor *(above)*. Clean out food deposits inside the spindle shaft with a small bottle brush. If the shaft is broken, or the ridges inside are worn or damaged, replace the shaft with an identical part from the manufacturer or an authorized service center. Reinstall the flat washer, drive wheel and lock washer. Reinstall the drive belt and adjust its tension *(step 4)*.

4 **Adjusting the drive belt.** Rotate the drive belt by hand to check that it fits snugly and moves smoothly with the motor pulley and drive wheel. If the belt slips off while turning, or turns with difficulty, adjust the tension. Use a nut driver to loosen the screw securing the bearing bracket. Slide the bracket to tighten or loosen the belt *(above)*, then screw the bracket firmly in its new position. Reinstall the lower housing and try the food processor. Readjust the belt if necessary.

TESTING AND REPLACING MULTI-CONTROL SWITCHES (Belt-drive food processors)

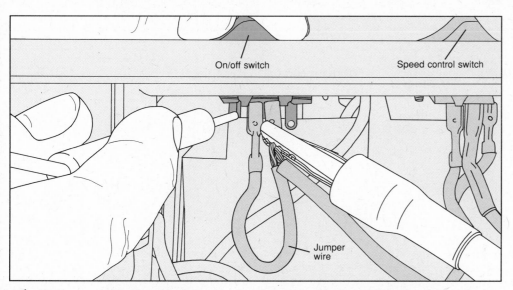

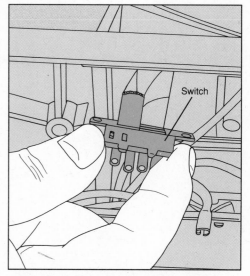

1 **Testing a multi-control switch.** The on/off and speed control switches are located in the upper housing. To gain access to them, unplug the food processor and remove the lower housing *(page 82)*. Test each switch separately. Push the switch to one of the ON positions. Use a continuity tester, or set a multitester to test continuity *(page 113)*. Clip one probe to the switch's terminal connected to a power cord wire. (If this terminal is attached to a jumper wire, as shown, do not disconnect the jumper wire.) Label and disconnect the other wires. Touch the other tester probe to each of the other two terminals, in turn *(above)*. The tester should show continuity at one terminal only. Set the switch to its other ON position and repeat the test. The tester should show continuity at the other terminal only. If the switch fails either test, replace it *(step 2)*.

2 **Replacing a multi-control switch.** Label and disconnect the power cord wire. Pull off the switch cap and peel back the nameplate to expose the switch screws. Unscrew and pull out the faulty switch *(above)* and install an exact replacement purchased from an authorized service center. Reconnect the wires, reinstall the lower housing and cold check for leaking voltage *(page 114)*.

TESTING AND REPLACING THE SAFETY SWITCH (Belt-drive food processors)

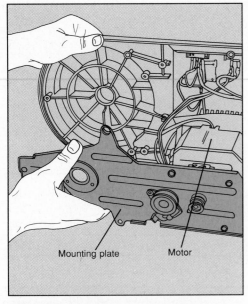

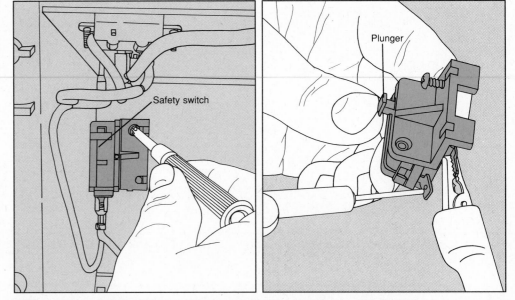

Mounting plate   Motor

Safety switch

Plunger

1 **Removing the mounting plate.** Unplug the food processor and take off the lower housing and drive assembly *(page 82)*. To gain access to the safety switch, remove the screws securing the mounting plate to the upper housing, then lift out the plate and motor *(above)* and slide it to one side, taking care not to damage or pull any wires.

2 **Testing and replacing the safety switch.** Locate the safety switch, deep inside the upper housing. Unscrew it from the housing wall *(above, left)* and lift it out. Label and disconnect the two wires. Use a continuity tester, or set a multitester to test continuity *(page 113)*. Touch a probe to each terminal; the tester should show no continuity. Then depress the plunger on the side of the switch *(above, right)*; the tester should show continuity. If the switch fails either test, replace it with an exact replacement, purchased from an authorized service center. Reassemble the food processor, making sure all wires are properly reconnected, and cold check for leaking voltage *(page 114)*.

DIRECT-DRIVE FOOD PROCESSORS

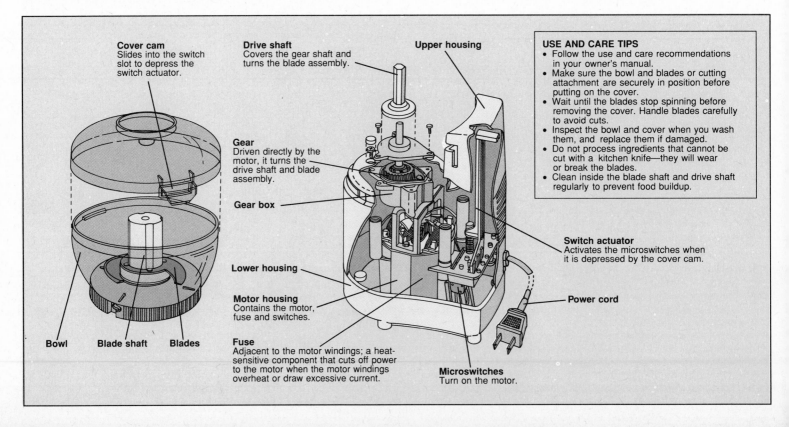

Cover cam
Slides into the switch slot to depress the switch actuator.

Drive shaft
Covers the gear shaft and turns the blade assembly.

Upper housing

Gear
Driven directly by the motor, it turns the drive shaft and blade assembly.

Gear box

Lower housing

Motor housing
Contains the motor, fuse and switches.

Bowl   **Blade shaft**   **Blades**

Fuse
Adjacent to the motor windings; a heat-sensitive component that cuts off power to the motor when the motor windings overheat or draw excessive current.

Switch actuator
Activates the microswitches when it is depressed by the cover cam.

Power cord

Microswitches
Turn on the motor.

USE AND CARE TIPS
- Follow the use and care recommendations in your owner's manual.
- Make sure the bowl and blades or cutting attachment are securely in position before putting on the cover.
- Wait until the blades stop spinning before removing the cover. Handle blades carefully to avoid cuts.
- Inspect the bowl and cover when you wash them, and replace them if damaged.
- Do not process ingredients that cannot be cut with a kitchen knife—they will wear or break the blades.
- Clean inside the blade shaft and drive shaft regularly to prevent food buildup.

SERVICING THE INTERNAL COMPONENTS (Direct-drive food processors)

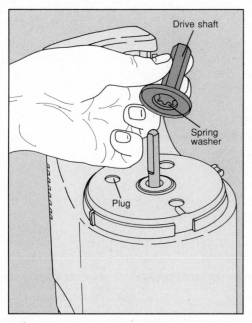

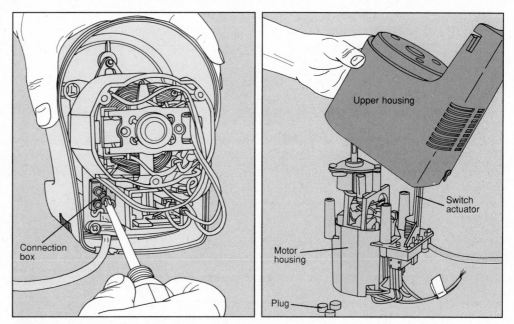

1 **Servicing the drive shaft.** Unplug the unit and remove the bowl and cover. If the drive shaft can be spun freely by hand or if its edges are worn, replace it with an exact duplicate from an authorized service center. Use a screwdriver to pry up the shaft and pull it off, with its spring washer attached *(above)*. Clean the drive shaft of any hardened food deposits.

2 **Inspecting the motor-and-gear assembly.** Unscrew the lower housing and pull it off. Remove the screw securing the connection box *(above, left)* and lift it up. Label and disconnect the wires attached to the box (one leads to the fuse), then remove the four screws recessed in the motor housing. Turn the unit upright and locate the plastic plugs in the top of the upper housing; they conceal the housing screws. Use a fine screwdriver to pry out the plugs, then unscrew and lift off the upper housing *(above, right)*. You now may service the gear assembly *(step 3)* or fuse *(step 4)*. If the motor smells burned or if the motor windings are charred, take the unit for professional service.

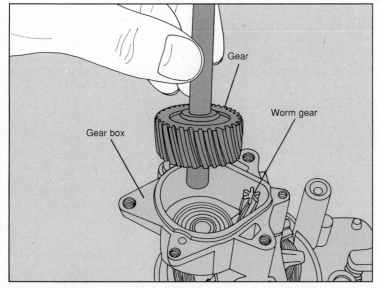

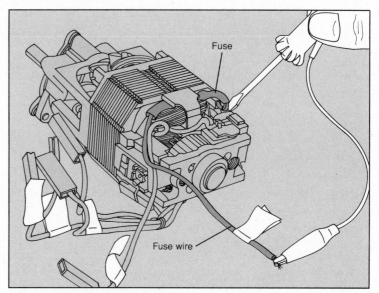

3 **Replacing the gear.** Unscrew and remove the gear box cover and lift out the gear *(above)*. If its teeth are worn or broken, buy an exact replacement gear from the manufacturer or an authorized service center. Before installing it, clean the worm gear with a small brush, and wipe old lubricant and debris from inside the gear box with a cloth. Lubricate the teeth of the new gear with a high-temperature multipurpose grease. Seat the gear, lubricate its top, and reinstall the cover. To test the fuse, go to step 4. Otherwise, reassemble the food processor, correctly reconnecting all wires, and cold check for leaking voltage *(page 114)*.

4 **Testing and replacing the fuse.** Lift out the switch actuator, then label and disconnect the wires to the switches. Slide the motor out of its housing. Follow the fuse wire to locate the fuse, near the motor windings, wrapped in a protective covering. Touch a continuity tester probe to each fuse-wire end *(above)*. The tester should show continuity. If the fuse tests OK, service the motor *(page 123)*. Otherwise, disconnect and replace the fuse. Buy an exact replacement fuse from an authorized service center and crimp on a new connector *(page 118)*. Install the fuse and reassemble the food processor, making sure all wires are properly connected. Cold check for leaking voltage *(page 114)*.

HUMIDIFIERS

By adding moisture to the air, humidifiers help keep furniture from warping, wallpaper from peeling and plaster from cracking during the dry winter months. Moist air also prevents dry skin, respiratory infection and even static cling.

The console-style evaporative humidifiers pictured below are two of the most common models. Each uses a fan to blow dry air through a moist rotating filter mounted on rollers or a drum. The filter may be driven by a variety of pulleys, drive belts, discs or gears, turned by the same motor that runs the fan. Because a humidifier runs almost continuously, its switches and motor suffer less wear and tear than those of

many other small appliances. When components do break down you can prolong the humidifier's life by fixing most problems yourself, consulting the Troubleshooting Guide at right. If access to the faulty part differs from what is shown in this chapter, consult Tools & Techniques (page 110) for tips on disassembly. However, if your humidifier is still under warranty, take it to an authorized service center.

In contrast to an evaporative humidifier, the ultrasonic humidifier works by breaking up molecules of water electronically and releasing them into the room as a mist. It is pictured and described on page 92.

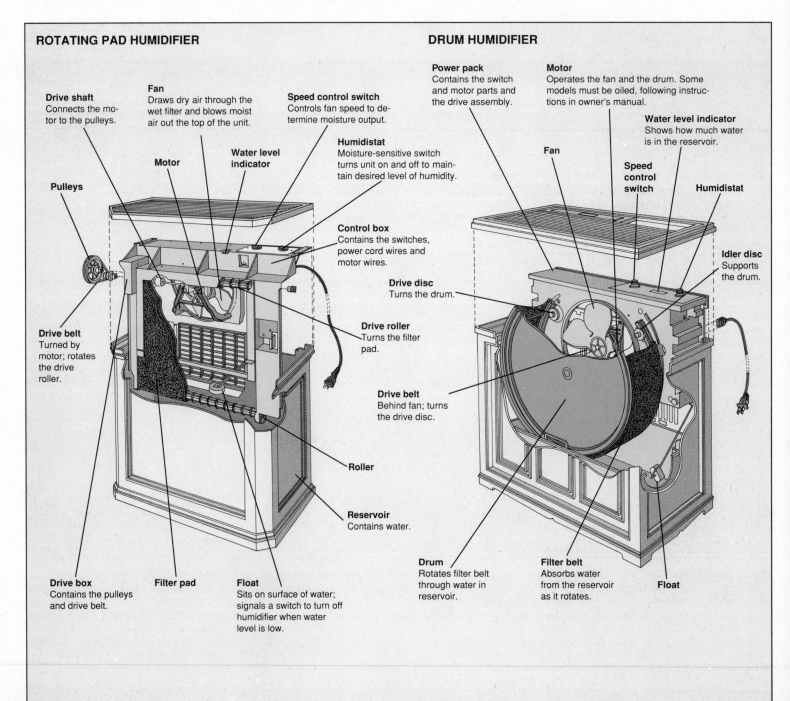

ROTATING PAD HUMIDIFIER

Drive shaft
Connects the motor to the pulleys.

Fan
Draws dry air through the wet filter and blows moist air out the top of the unit.

Speed control switch
Controls fan speed to determine moisture output.

Humidistat
Moisture-sensitive switch turns unit on and off to maintain desired level of humidity.

Motor

Water level indicator

Pulleys

Control box
Contains the switches, power cord wires and motor wires.

Drive belt
Turned by motor; rotates the drive roller.

Drive roller
Turns the filter pad.

Roller

Reservoir
Contains water.

Drive box
Contains the pulleys and drive belt.

Filter pad

Float
Sits on surface of water; signals a switch to turn off humidifier when water level is low.

DRUM HUMIDIFIER

Power pack
Contains the switch and motor parts and the drive assembly.

Motor
Operates the fan and the drum. Some models must be oiled, following instructions in owner's manual.

Water level indicator
Shows how much water is in the reservoir.

Fan

Speed control switch

Humidistat

Idler disc
Supports the drum.

Drive disc
Turns the drum.

Drive belt
Behind fan; turns the drive disc.

Drum
Rotates filter belt through water in reservoir.

Filter belt
Absorbs water from the reservoir as it rotates.

Float

TROUBLESHOOTING GUIDE

SYMPTOM	POSSIBLE CAUSE	PROCEDURE
Humidifier does not run at all	Humidifier unplugged or turned off	Plug in and turn on humidifier
	No power to outlet	Reset breaker or replace fuse *(p. 112)* □○; have outlet serviced
	Power cord faulty	Test and replace power cord *(p. 116)* ◪○▲
	Humidistat set too low	Turn up humidistat
	No water in reservoir	Fill reservoir
	Float waterlogged	Replace float □○
	Float switch, humidistat or speed control switch faulty	Service switches *(p. 88)* ◪○
	Motor faulty	Service motor *(p. 90)* ◪○▲
Motor and fan run, but filter doesn't turn	Drive belt loose or broken	Service drive belt *(p. 91)* □○
	Discs or gears faulty	Service discs or gears *(p. 92)* □○
Humidifier is noisy	Discs or gears faulty	Service discs or gears *(p. 92)* □○
	Motor bearings dry	Oil motor bearings according to instructions in owner's manual
Humidifier doesn't humidify	Filter pad or filter belt dirty	Clean or replace filter pad or filter belt *(rotating pad type, p. 87; drum type, p. 88)* □○
Humidifier smells bad	Bacteria buildup in reservoir	Clean reservoir *(rotating pad type, p. 87; drum type, p. 88)* □○
Ultrasonic humidifier produces little or no mist	Transducer dirty or faulty	Service the humidifier *(p. 92)* □○

DEGREE OF DIFFICULTY: □ Easy ◪ Moderate ■ Complex
ESTIMATED TIME: ○ Less than 1 hour ◐ 1 to 3 hours ● Over 3 hours   ▲ Special tool required

CLEANING A FILTER PAD AND RESERVOIR (Rotating pad type)

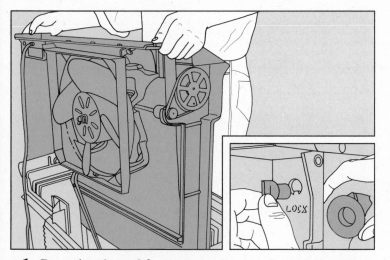

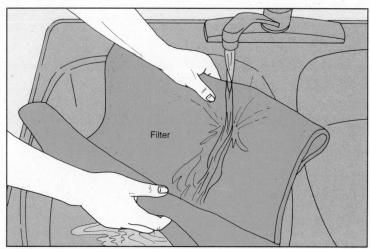

Filter

1 Removing the pad frame. Turn off and unplug the humidifier. Spread newspaper on the floor. Take off the lid, and remove and discard any shipping screws securing the pad frame to the reservoir. Slide the pad frame up out of the channels in the sides of the reservoir *(above)* and lay it on the newspaper. To remove the filter pad for cleaning or replacement, release the bottom roller by pulling the frame sides apart by hand. Then release the top roller by turning the roller bearing clockwise to the vertical position and pulling it free *(inset)*. Slide the pad off the rollers. Inspect the ends of the rollers. If they are worn or broken, replace them with new ones purchased at an authorized service center.

2 Cleaning the filter and reservoir. Soak the filter in a solution of equal parts vinegar and water, then rinse it under running water *(above)*. If stubborn mineral deposits remain, replace the filter with a new one from an authorized service center. Clean the reservoir by filling it with the vinegar solution for a couple of hours, then scrub it well with a stiff brush and rinse it out. To kill odors, wash the reservoir with detergent and water, then disinfect it with a solution of one-half cup bleach per gallon of water. Rinse the reservoir again. Fit the pad on the rollers and install the rollers in the frame. Reassemble the humidifier.

CLEANING A FILTER BELT AND RESERVOIR (Drum type)

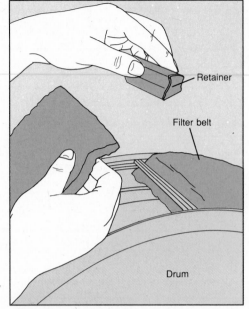

1 **Removing the drum.** Spread out newspaper to protect the floor. Turn off and unplug the humidifier and take off its lid. Lift the drum up off the discs on which it hangs, pull it forward slightly, then raise it straight up out of the humidifier *(above)*. Lay the drum on the newspaper.

2 **Removing the filter belt.** Pry out the plastic filter belt retainer between the ribs of the drum and release the ends of the filter *(above)*. If there is no retainer, slip the belt off the drum in one piece. Wash the filter belt *(page 87, step 2)*. Fit the belt on the drum while it is still wet, reversing the steps you took to remove it. To clean the reservoir, go to step 3.

3 **Removing the power pack.** If the power cord is detachable, remove it. Unscrew and discard any shipping screws that secure the power pack to the chassis. Pull the power pack up and out of the humidifier *(above)*. Lift out the reservoir carefully; it may still contain water. Wash the reservoir thoroughly *(page 87, step 2)*. Reinstall the reservoir, power pack and drum, reversing these steps.

TESTING AND REPLACING SWITCHES

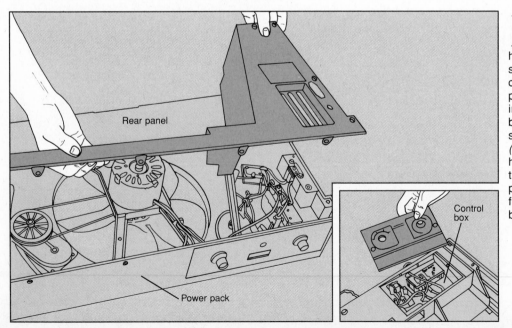

1 **Accessing the switches.** All evaporative humidifiers are equipped with a float switch and a humidistat. Many also have a speed control switch. To access the switches for testing and replacement, turn off and unplug the humidifier. On a rotating pad type humidifier such as the one shown in this chapter, simply pry off the control buttons with your fingers. Then remove the screws from the control panel and lift it off *(inset)*. On the drum type model shown here, you must first remove the drum and the power pack *(steps above)*. Lay the power pack on a sturdy flat surface with the fan facing down. Unscrew and remove the back panel *(left)* and lay it aside.

TESTING AND REPLACING SWITCHES (continued)

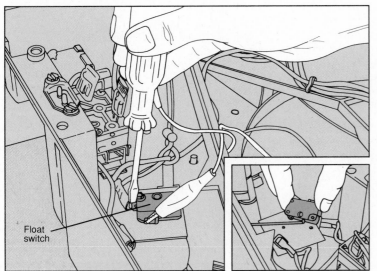

Float switch

2 **Testing and replacing a float switch.** Identify the float switch by the metal lever attached to its side. Disconnect one wire from the switch. Lower the float arm as far as it will go and touch a continuity tester probe to each terminal *(left)*. The tester should indicate no continuity. Raise the float arm slowly and listen for a click of the switch. The tester should indicate continuity after the switch clicks. If the switch tests OK, go to step 3 and test the humidistat. If the switch fails this test, replace it. Pull off any remaining wire and unscrew and remove the switch, noting how the metal lever makes contact with the float arm for reassembly *(inset)*. Buy an exact duplicate switch from an authorized service center and install it, using the same screws you removed. Reconnect the wires and reassemble the humidifier, reversing the steps you took to access the switches *(step 1)*.

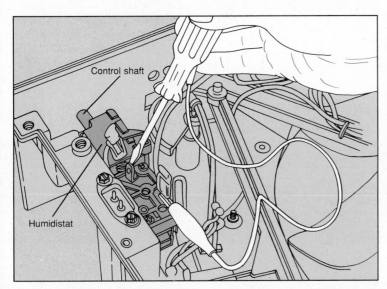

Control shaft

Humidistat

3 **Testing and replacing a humidistat.** The humidistat is a large rectangular switch with a control shaft protruding from it. Turn the control counterclockwise to the OFF position. Disconnect one wire from the humidistat and touch a continuity tester probe to each terminal *(left)*. The tester should indicate no continuity. Turn the control slowly toward ON and listen for a click. The tester should indicate continuity after the humidistat clicks. If the humidistat tests OK, go to step 4 and test the speed control switch. If the humidistat tests faulty, replace it with an exact duplicate purchased from an authorized service center. Pull off any remaining wire. Pry the humidistat control knob off the control panel, if you haven't already. Remove all screws securing the humidistat and lift it out. Install the new humidistat in the same position as the old one. Reconnect the wires and reassemble the humidifier, reversing the steps you took to access the switches *(step 1)*.

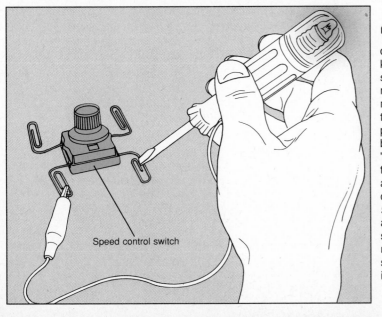

Speed control switch

4 **Testing and replacing a speed control switch.** The small, square speed control switch has a rotary control shaft at its center. Unscrew the switch and take it out to test it: On the drum humidifier shown here, you must pry off the speed control knob, then cut away a bit of the metal panel around the control shaft to access the mounting screws *(page 115)*. Label and disconnect the wires. Most speed control switches have self-locking terminals. To remove a wire, insert the tip of a tiny screwdriver into the terminal hole and pull out the wire. Stick a partly-straightened paper clip into each terminal that held a wire. Put the control knob back on the shaft. Touch the clip in the common terminal (marked "L" or "C") with one continuity tester probe, while you touch each of the other clips in turn with the second probe *(left)*. Repeat this test at each speed setting. The tester should indicate continuity at least once for each setting. If the switch tests OK, test the motor *(page 90)*. If the switch tests faulty, purchase an exact replacement from an authorized service center. Place the control knob on the control shaft and orient the switch so that the knob indicator matches the markings on the control panel. Then remove the knob, install the switch and reconnect the wires. Reassemble the humidifier, reversing the steps you took to access the switches *(step 1)*.

TESTING AND REPLACING THE MOTOR

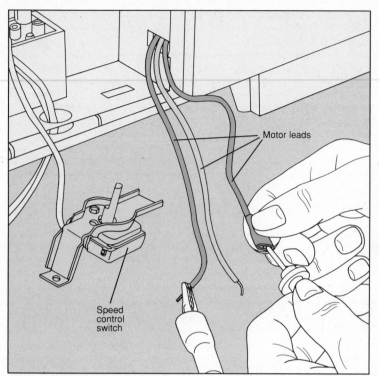

Motor leads

Speed control switch

1 **Testing the motor.** Most humidifiers have multi-speed motors, with three or four wire leads. Before testing the motor, turn off and unplug the humidifier and take off the lid. To locate the motor leads on the humidifier models shown here, access the switches *(page 88, step 1)*. Remove the speed control switch, and identify and disconnect the motor leads from its terminals *(page 89, step 4)*. Leave the power cord wire attached. Disconnect the remaining motor lead, usually white, from the float switch. Set a multitester to RX1 *(page 113)*. With a probe clipped to the end of the motor lead from the float switch, touch the second probe to the end of each other motor lead in turn *(left)*. The multitester should indicate partial resistance for each test. If the motor tests OK, inspect the wire connections and repair them *(page 118)*. If the motor has infinite resistance, or has continuity, for any test, replace it: For the motor in a rotating pad type humidifier, go to step 2; in a drum type humidifier, go to step 3.

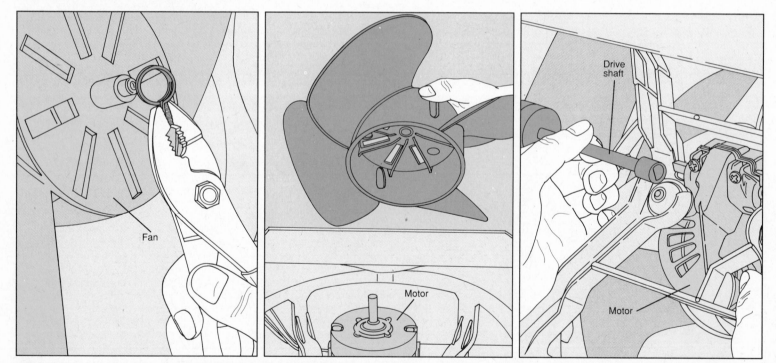

Fan

Motor

Drive shaft

Motor

2 **Replacing the motor (rotating pad type).** The fan on a rotating pad type humidifier usually must be removed before you can take out the motor. On the model shown here, the fan is secured to the motor shaft by a retainer ring and a spring clamp. Position the pad frame so that the fan is facing up. Pry off the ring with a screwdriver, then squeeze the spring clamp with pliers and remove it *(above, left)*. Lift the fan off the shaft *(above, center)*. Turn over the pad frame and take off the filter pad and rollers *(page 87)*. Pull the drive shaft away from the motor shaft to disengage it *(above, right)*; if the foam washer falls out, stick it back into the shaft. Unscrew the nuts holding the motor, noting the positions of the washers for reassembly. Lift off the pad frame, pulling the motor leads free. Take the motor to an authorized service center and purchase an exact replacement, or order one from the manufacturer. To install the new motor, reverse these steps, supporting the motor with one hand as you tighten the motor mounting nuts with the other. Reinstall the drive shaft, aligning its flat side with the flat side of the motor shaft. Reconnect the motor leads to their correct switch terminals, using the labels for reference. Remount the speed control switch and reassemble the humidifier.

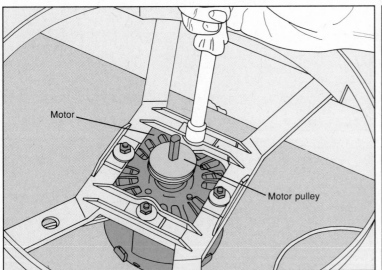

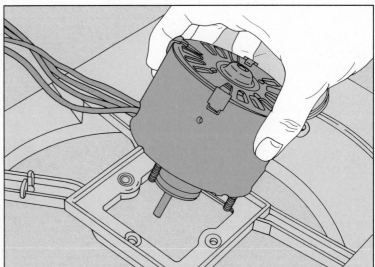

3 **Replacing the motor (drum type).** Turn the power pack so that the fan is facing up. Pull the fan off the end of the motor shaft. Take off the drive belt. Unscrew the mounting nuts that secure the motor to the mounting bracket on the power pack *(above, left)*. Keep track of the washers for reassembly. Lift the power pack off the motor. Pry the motor pulley off the motor shaft with a blunt tool, such as a pair of pliers, taped to protect the pulley. Buy an exact replacement motor from an authorized

service center or the manufacturer. To install the new motor, place it in the power pack mounting bracket *(above, right)*. Turn over the power pack, holding the motor in place, and screw on the nuts and washers. Push the motor pulley onto the motor shaft and slip on the drive belt. Push the fan onto the end of the motor shaft, aligning its flat side with the flat side of the shaft. Turn over the power pack and reconnect the motor leads to their correct switch terminals. Reinstall the speed control switch and reassemble the humidifier.

SERVICING THE DRIVE BELT

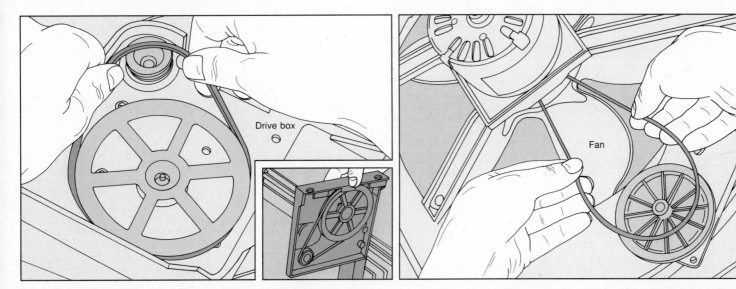

Replacing a drive belt. Turn off and unplug the humidifier, and take off the lid. The drive belt may be accessed in several ways, depending on the model. Some humidifiers have two belts, connecting the motor to the drive pulley and to the fan. On the rotating pad type shown here, you must first take out the pad frame *(page 87, step 1)*. On other types, after removing the pad assembly, you may have to disengage the transmission shaft *(page 90, step 2)* before you can unscrew and lift out the drive box containing the drive belt *(inset)*. On drum type humidifiers, you must remove the drum and the

power pack *(page 88)* to reach the drive belt. Inspect the belt for wear or stretching. Once you have access to the drive belt on most humidifiers, simply slip it off its pulleys *(above, left)*. In the drum type shown here, however, you must also stretch the belt around the fan blades to take it out of the humidifier *(above, right)*. Purchase an exact replacement belt from a small appliance parts dealer. Install the new belt by reversing the steps you took to remove the old one, then reassemble the humidifier.

SERVICING DISCS AND GEARS (Drum type)

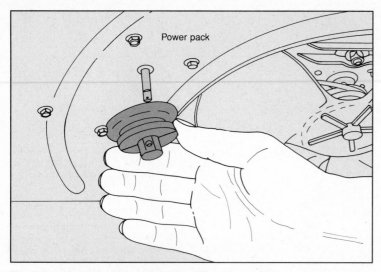

Servicing the drive disc and idler disc. To inspect or replace drive and idler discs, turn off and unplug the humidifier. Take off the lid and lift out the drum *(page 88)*. To remove the drive disc from the power pack, unscrew the retaining screw and pull the disc off the axle *(above)*. To remove the idler disc, pull off its plastic cap with pliers, then unscrew it and slide it off. To dissolve mineral deposits, soak the discs in a half-and-half solution of vinegar and water, then scrub them with a stiff brush. If a disc is worn or broken, purchase a replacement from an authorized service representative. Mount the new disc securely on its axle. Reinstall the drum and lid.

Servicing drive gears. The drum on some humidifiers is turned by a drive gear whose teeth mesh with a gear on the drum. Turn off and unplug the humidifier. Take off the lid and remove the drum *(page 88)*. Inspect the drum teeth for wear and mineral buildup. Unscrew the retaining bracket that covers the front of the drive gear assembly and slide the drive gears off the gear shafts *(above)*. Soak the drive gears in a half-and-half solution of vinegar and water and clean them with a stiff brush *(inset)*. If the gear teeth are worn or broken, buy replacement gears from an authorized service representative. Slip the gears back onto their shafts, replace the retaining bracket and put back the drum and lid.

SERVICING AN ULTRASONIC HUMIDIFIER

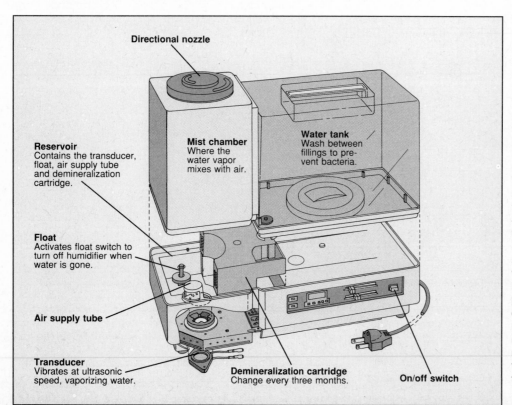

Directional nozzle

Mist chamber
Where the water vapor mixes with air.

Water tank
Wash between fillings to prevent bacteria.

Reservoir
Contains the transducer, float, air supply tube and demineralization cartridge.

Float
Activates float switch to turn off humidifier when water is gone.

Air supply tube

Transducer
Vibrates at ultrasonic speed, vaporizing water.

Demineralization cartridge
Change every three months.

On/off switch

Troubleshooting an ultrasonic humidifier. This electronic device has a small oscillating disc, called a transducer, that vibrates water at an ultrasonic frequency, breaking apart the water molecules to create a fine mist. In the process, however, minerals in hard water may precipitate, appearing as white dust on surrounding surfaces and as stubborn mineral deposits inside the humidifier itself. To prevent this, consult the use and care recommendations in the owner's manual, or use distilled water. Keep an ultrasonic humidifier away from sensitive electronic equipment such as stereos and televisions and never turn it on when empty. Ultrasonic humidifiers rarely break down. If yours stops working, first check its power cord *(page 116)*. If air is emitted from the nozzle but no mist, make sure that the water tank is filled with tepid water and that the humidistat control is set high enough. If the humidifier still doesn't produce mist, suspect the transducer, and clean it or test and replace it *(page 93)*. On a model such as the one shown here, you can install a new transducer at home. For other models, consult a humidifier service center first. Circuit board replacement or recalibration are best performed by a professional.

SERVICING AN ULTRASONIC HUMIDIFIER (continued)

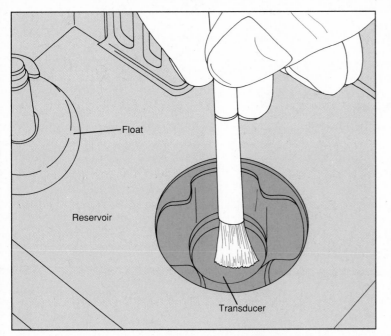

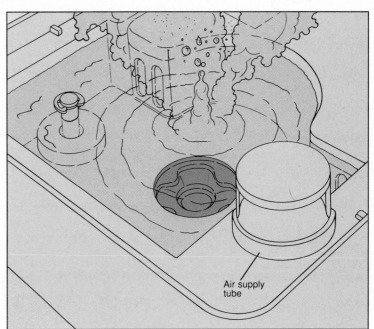

1 **Cleaning the reservoir, float and transducer.** Turn off and unplug the unit, and remove the mist chamber and water tank. Empty the reservoir. Do not use detergent; use a soft damp cloth to wipe the reservoir, and clean the float and transducer with nothing harder than a soft-bristled brush dipped in vinegar or commercial scale remover *(above)*. Avoid putting strong pressure on the transducer, and avoid touching the transducer with your fingers; oily residues reduce its efficiency. To dissolve stubborn deposits, fill the reservoir with vinegar and let it stand for 30 minutes. Empty the reservoir and brush the transducer until it is shiny. Rinse well. If the humidifier still doesn't produce mist after cleaning, go to step 2 to test the transducer.

2 **Testing the transducer.** To check that the transducer is working, fill the reservoir with water, plug in the unit, set the humidistat and mist intensity controls to their highest position and turn on the power. Keep your hands away from the reservoir area. If the water in the reservoir agitates but no mist is produced, the transducer is faulty and must be replaced *(step 3)*. If the water in the reservoir doesn't agitate at all, suspect a faulty circuit board or other electrical component, and take the humidifier for professional service.

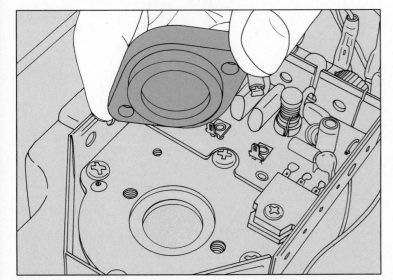

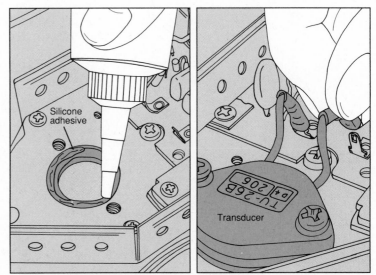

3 **Removing the transducer.** Turn off and unplug the humidifier. Empty the reservoir. Turn over the humidifier body and unscrew and remove the base plate. Pull both transducer wires off the circuit board. Unscrew the two mounting screws, keeping track of the small metal washers. Lift the transducer off the mounting plate *(above)*. Buy an exact replacement transducer from an authorized service center or the manufacturer.

4 **Replacing a transducer.** To install a new transducer, apply a continuous bead of silicone adhesive around the hole in the mounting plate *(above, left)*. Press the transducer down firmly, fitting its rubber ring into the hole. Screw the transducer to the mounting plate using the washers and mounting screws. Reconnect the transducer wires to their terminals on the circuit board *(above, right)*. Screw on the base plate and reassemble the unit. Allow 24 hours for the adhesive to dry before using the humidifier.

VACUUM CLEANERS

The indispensable vacuum cleaner had its humble beginnings as an electric fan motor, mounted in a soap box, that collected dirt in a pillow case. Since then, it has been streamlined and modified into two basic styles: upright and canister. Two typical models are pictured below.

A vacuum cleaner sucks up dirt and air, filters out the dirt and allows the air to escape. All vacuum cleaners use a motor and fan to generate this suction; what distinguishes the two types is the route taken by the air flow. In an upright vacuum cleaner, dirt dislodged by the beater bar is pulled by the airstream into the dirt fan, which whirls it up into the dust bag. In a canister model, dirt is sucked through a hose and trapped in the dust bag as it enters the canister; only clean air passes the

fan. A major cause of vacuum cleaner malfunction is, of course, dirt. Change the dust bag often. Periodically check the filter in a canister vacuum cleaner, and wash or replace it as necessary. Keep spare dust bags and filters on hand.

Attentive maintenance will prevent many vacuum cleaner problems, but parts do wear out, most typically the drive belt and the beater bar brushes. If your vacuum cleaner does not operate quietly and efficiently, consult the Troubleshooting Guide at right. Vacuum cleaners are among the most easily serviced of household appliances. You can repair or replace most parts yourself, including the motor brushes and bearing. Canister types, especially older models, may differ from the ones shown here. Use this chapter as a general guide to repair.

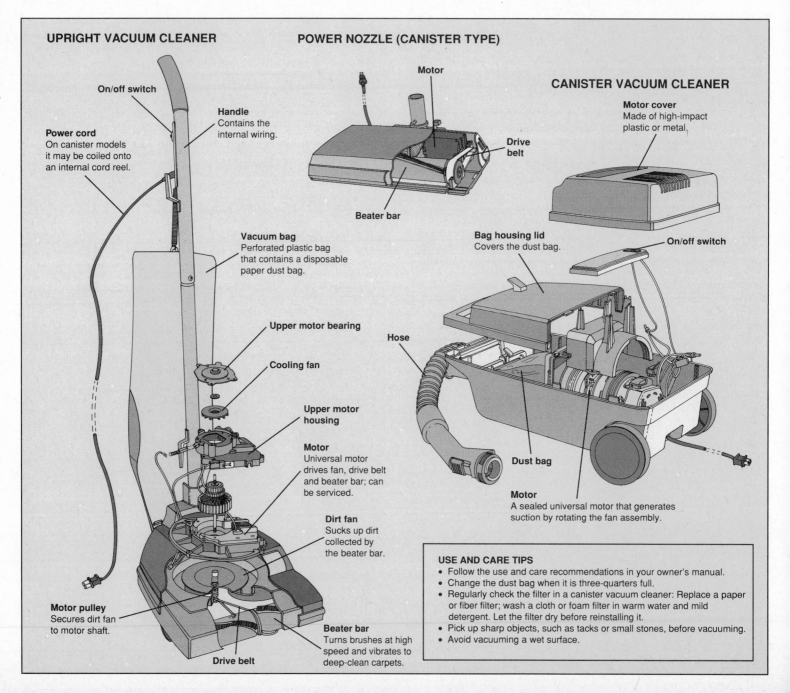

UPRIGHT VACUUM CLEANER

On/off switch

Power cord
On canister models it may be coiled onto an internal cord reel.

Handle
Contains the internal wiring.

Vacuum bag
Perforated plastic bag that contains a disposable paper dust bag.

Upper motor bearing

Cooling fan

Upper motor housing

Motor
Universal motor drives fan, drive belt and beater bar; can be serviced.

Dirt fan
Sucks up dirt collected by the beater bar.

Motor pulley
Secures dirt fan to motor shaft.

Drive belt

Beater bar
Turns brushes at high speed and vibrates to deep-clean carpets.

POWER NOZZLE (CANISTER TYPE)

Motor

Drive belt

Beater bar

CANISTER VACUUM CLEANER

Motor cover
Made of high-impact plastic or metal.

Bag housing lid
Covers the dust bag.

On/off switch

Hose

Dust bag

Motor
A sealed universal motor that generates suction by rotating the fan assembly.

USE AND CARE TIPS
- Follow the use and care recommendations in your owner's manual.
- Change the dust bag when it is three-quarters full.
- Regularly check the filter in a canister vacuum cleaner: Replace a paper or fiber filter; wash a cloth or foam filter in warm water and mild detergent. Let the filter dry before reinstalling it.
- Pick up sharp objects, such as tacks or small stones, before vacuuming.
- Avoid vacuuming a wet surface.

TROUBLESHOOTING GUIDE

SYMPTOM	POSSIBLE CAUSE	PROCEDURE
Vacuum cleaner doesn't turn on	Vacuum cleaner unplugged or turned off	Plug in and turn on vacuum cleaner
	No power to outlet or outlet faulty	Reset circuit breaker or replace fuse (p. 112) □○; have outlet serviced
	Power cord faulty	Test and replace power cord (p. 116) ▙○▲; service cord reel assembly (canister type with cord winder, p. 98) ▙○
	On/off switch faulty	Test and replace on/off switch (p. 100) ▙○▲
	Wire connections loose or broken	Inspect and test wiring (p. 103) ▙○
	Motor faulty	Service motor (p. 103) ▙◑▲
Vacuum cleaner turns on, then stops	Household electrical circuit overloaded	Reduce the number of appliances on circuit
	Motor overheated due to full dust bag or dirty filter	Turn off vacuum cleaner, replace dust bag and clean or replace filter; wait for motor to cool and thermal protector to reset itself before turning on vacuum cleaner
Vacuum cleaner runs intermittently	Power cord faulty	Test and replace power cord (p. 116) ▙○▲; service cord reel assembly (canister type with cord winder, p. 98) ▙○
	On/off switch faulty	Test and replace on/off switch (p. 100) ▙○▲
	Wire connections loose or broken	Inspect and test wiring (p. 103) ▙○
	Motor faulty	Service motor (p. 103) ▙◑▲
Vacuum cleaner hums, smokes or overheats	Fan jammed or dirty	Service fan (upright type, p. 101; canister type, p. 102) ▙○
	Motor brushes worn or field coil shorted	Service motor (p. 103) ▙◑▲
	Motor shaft bearing binding	Service motor (upright type, p. 103) ▙◑▲; take canister type for professional service
Vacuum cleaner doesn't clean any surface	Dust bag too full	Replace dust bag
	Filter dirty or blocked (canister type)	Clean or replace filter
	Hose clogged by dirt or foreign object (canister type)	Unblock hose with a broom handle
	Puncture in hose (canister type)	Tape or replace hose
Vacuum cleaner doesn't deep-clean carpets	Beater bar brushes clogged by debris (upright type and power nozzles)	Remove lint, hair and debris from beater bar (p. 97) □○
	Beater bar set too low or high (upright type)	Consult owner's manual and adjust wheel height
	Beater bar brushes worn (upright type and power nozzles)	Replace beater brushes (p. 97) ▙○
Beater bar rotates poorly or not at all (upright type and power nozzles)	Drive belt loose or broken	Replace drive belt (p. 97) □○
	Beater bar bearings jammed with dirt or lint	Clean and lubricate bearings (p. 97) □○
	Pressure clip holding beater bar broken	Replace broken clip (p. 97) □○
	Wire connection between vacuum cleaner and power nozzle loose or broken	Repair wire connection (p. 103) □○
Vacuum cleaners is noisy or vibrates excessively	Foreign object caught in fan; fan loose or damaged	Service fan (upright type, p. 101; canister type, p. 102) ▙○
	Drive belt worn or broken (upright type and power nozzles)	Replace drive belt (p. 97) □○
	Beater bar loose (upright type and power nozzles)	Replace beater-bar pressure clips (p. 97) □○
	Motor shaft bearing worn	Service motor (upright type, p. 103) ▙◑▲; take canister type for professional service
Power cord does not wind onto cord reel (canister type with cord winder)	Main spring in cord reel broken	Replace cord reel (p. 98) ▙○

DEGREE OF DIFFICULTY: □ Easy ▙ Moderate ■ Complex
ESTIMATED TIME: ○ Less than 1 hour ◑ 1 to 3 hours ● Over 3 hours   ▲ Special tool required

ACCESS TO INTERNAL PARTS (Upright type)

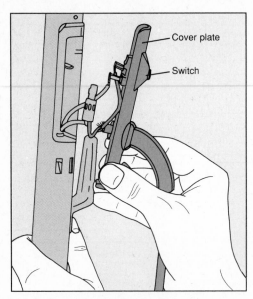

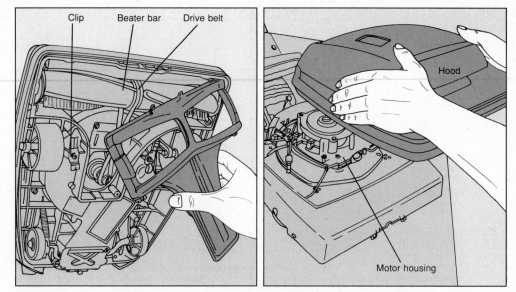

Removing and reinstalling the switch cover plate. Unplug the vacuum cleaner. Lock the handle in the upright position and unhook the vacuum cleaner bag. Unscrew the cover plate and slide it up off the handle *(above)*. After completing repair to the power cord or on/off switch, reinstall the cover plate and cold check for leaking voltage *(page 114)*.

Removing and reinstalling the bottom plate and the hood. Unplug the vacuum cleaner. Lock the handle in the upright position and lower it to the ground. Remove the bottom-plate mounting screws, or pivot the two retaining clips, then swing up the plate and pull it free *(above, left)*. You now have access to the drive belt and the beater bar. To repair the fan or motor, release the hood by unscrewing its mounting screws from the bottom. Then stand the vacuum cleaner back up and lower the handle to its lowest position. Lift off the hood *(above, right)*. After completing repairs, reinstall the hood and bottom plate and cold check for leaking voltage *(page 114)*.

ACCESS TO INTERNAL PARTS (Canister type)

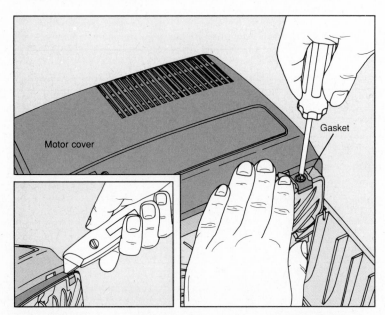

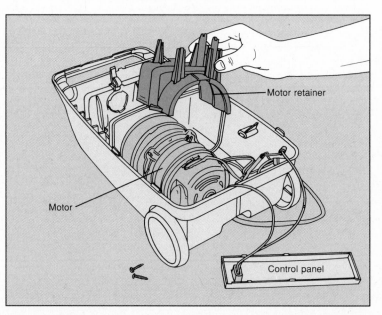

Removing and reinstalling the motor cover, control panel and motor retainer. Unplug the vacuum cleaner and protect the floor with old newspaper. Remove the bag housing lid and the dust bag. Locate the two screws that secure the motor cover to the filter support; rubber gaskets may conceal them. If so, use a utility knife to slit the gaskets *(inset)*, then peel away enough to reach the screw heads with a screwdriver. Remove the screws *(above, left)* and lift off the motor cover. The control panel on some canister models

comes off with the motor cover. On other models, such as the one shown above, it rests on the motor retainer and can be lifted off after the cover is removed. If the motor is shielded by a foam-and-cardboard muffler, remove it. Locate the screws that secure the motor retainer to the chassis, unscrew them and remove the retainer *(above, right)*. When repairs are complete, reinstall each part in order. Check that all wire connections are sound and cold check for leaking voltage *(page 114)*.

REPLACING THE DRIVE BELT (Upright type and power nozzles)

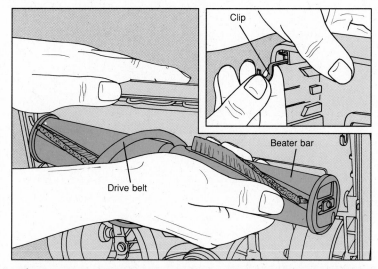

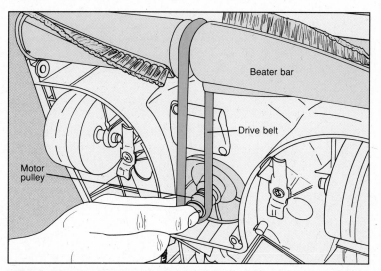

1 **Removing the drive belt.** Unplug the vacuum cleaner and remove the bottom plate *(page 96)*. Inspect the drive belt; if the belt is broken, check the owner's manual for its correct position. If the belt is damaged, note how it is installed for reference. On an upright vacuum cleaner, unhook the belt from the motor pulley and pull the beater bar free of its clips *(above)*. On a power nozzle, lift off the retaining bracket, pull out the beater bar and then unhook the belt from the motor drive shaft. Sometimes the clips that hold the beater bar become loose. Use pliers to pull them out and buy exact replacements at a vacuum cleaner service center. Fit a new clip back into its slot with your fingers *(inset)*.

2 **Replacing the drive belt.** Purchase an exact replacement belt at a vacuum cleaner service center. On the upright type, slide the belt onto its groove on the beater bar, snap the bar back into place and twist the belt around the motor pulley *(above)*. To reinstall a belt on a power nozzle, hook it around the drive shaft and then around its groove on the beater bar. Using both hands for added force, push the beater bar into its molded slots in the nozzle chassis, being careful not to pinch your fingers.

SERVICING THE BEATER BRUSH ASSEMBLY (Upright type and power nozzles)

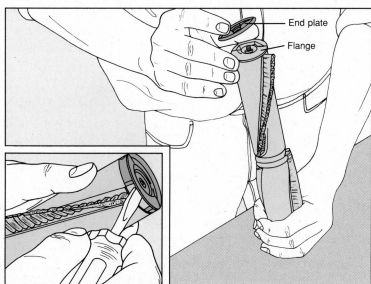

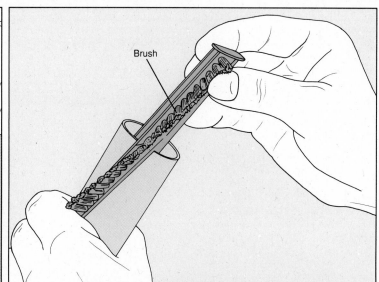

Disassembling the beater bar. Unplug the vacuum cleaner. Remove the bottom plate *(page 96)* and disengage the drive belt before pulling out the beater bar *(step above)*. Clean tangled hair, string and other debris from the bar and lubricate the beater bearings with graphite. To remove worn brushes, hold one end of the bar in each hand and turn them in opposite directions; one end plate will come off *(above, left)* revealing an inner metal flange and shaft. Pull off the other end plate; in some cases, you may have to tap the threaded end of the shaft with a hammer until the plate drops

off the beater bar. Pry off the inner metal flange with a screwdriver *(inset)*. Replace both brushes if either is worn. To remove a brush, grasp one end and pull it out of the slot in the beater bar *(above, right)*. Buy exact replacement brushes from a vacuum cleaner service center and slide them into their slots. Replace the inner flange, reinsert the threaded shaft and screw on the end plates tightly. Reinstall the drive belt and beater bar *(step above)* and the bottom plate *(page 96)*.

SERVICING THE CORD REEL ASSEMBLY (Canister type with cord winder)

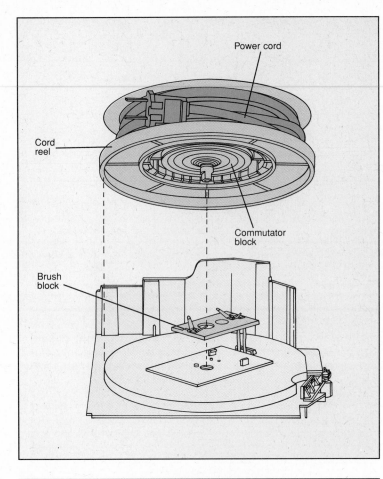

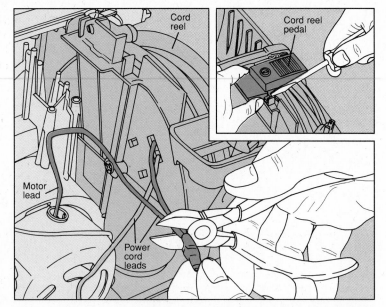

1 **Cutting the cord-reel assembly wires.** Unplug the vacuum cleaner and remove the motor cover and control panel *(page 96)*. Insert a screwdriver between the plastic clip retainers of the cord reel pedal and give the screwdriver a slight twist *(inset)* to release the pedal; note how the cord reel spring fits into a notch in its side for reference. Before removing the cord reel assembly, label and cut the power cord wires: Trace the two wires from where they exit the assembly; one is crimped to a motor lead and the other to the on/off switch leads. Label the motor lead and the on/off switch leads with masking tape and then cut off the crimp connectors with wire cutters *(above)*.

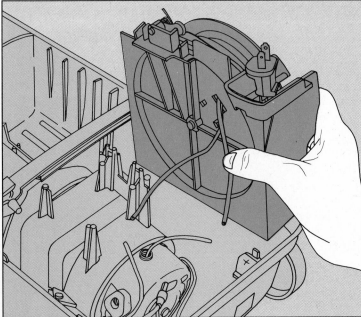

2 **Removing the cord reel assembly.** Note how the cord reel assembly sits in the canister cavity; the reel axle fits snugly into a socket in the cavity wall, and the reel support itself may be positioned in raised channels in the wall or in the back of the filter support. To free the assembly, pull it away from the side cavity wall and then push the filter support with one hand as you pull the reel toward you with the other. Lift it out of the cavity *(above)*.

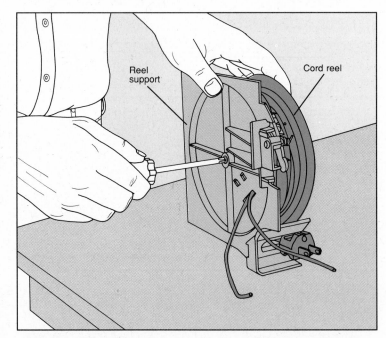

3 **Disassembling the cord reel.** Carefully unwind the cord reel: Grasp the power cord plug, free it from its slot, and let the power cord unwind slowly until it is no longer under tension. Set the assembly on end and remove the center fastening screw with a screwdriver *(above)*.

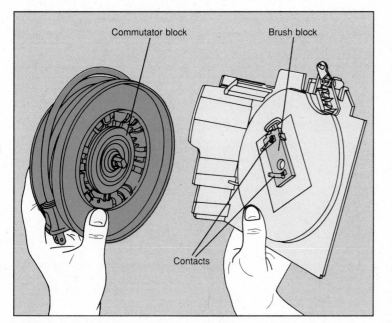

Commutator block

Brush block

Contacts

4 **Checking the brush block and commutator**. Pull apart the cord reel assembly *(above)*. If the main spring in the cord reel is broken, buy an exact replacement reel from a vacuum cleaner service center or the manufacturer and put the assembly back together *(step 6)*. If your vacuum cleaner has an electrical problem, examine the commutator block and the contacts on the brush block for carbon buildup and pitting. Wipe them clean with a paper towel or, if the carbon sticks, gently rub it off with a piece of fine emery paper. Lubricate the commutator lightly with petroleum jelly. If the contacts are pitted, replace the brush block *(step 5)*.

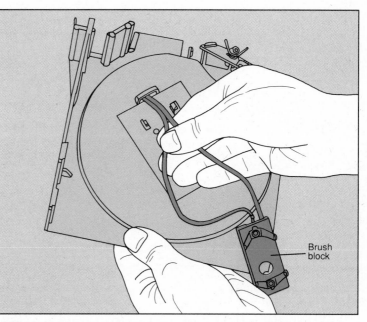

Brush block

5 **Replacing the brush block.** Remove the brush block from the plastic support by snapping it free of the clips with a screwdriver. Buy an exact replacement block from a vacuum cleaner service center or the manufacturer. To install a new brush block, feed the wire leads of the new block through the hole on the inside of the support *(above)* and snap the brush block into position. To ensure adequate contact between the contacts and the commutator block, use pliers to bend the contacts outward slightly.

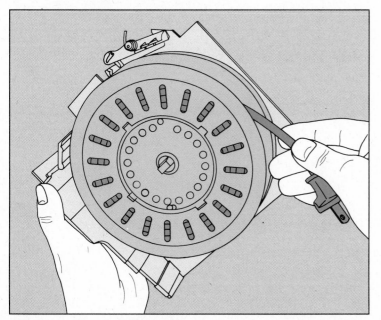

6 **Rewinding the cord reel.** Fit the cord reel into its support and reinstall the screw in the center. Hold the cord reel in one hand so that it faces you, grasp the plug with the other hand and wind the power cord around the reel four times in a clockwise direction; do not overwind the cord or the main reel spring may break. Reposition the power cord plug in its slot to keep the power cord from unwinding.

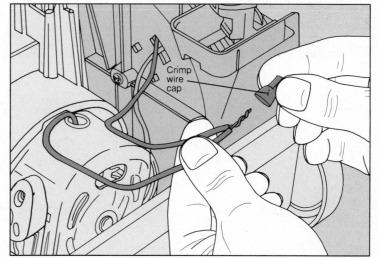

Crimp wire cap

7 **Crimping the wires.** Install the cord reel assembly in the canister cavity, inserting the reel axle securely in its socket and positioning the reel support correctly in its channels. Strip about 1/2 inch of insulation from the ends of all the cut wires with wire strippers *(page 118)*. Twist together the motor lead and one of the power cord wires and slide a crimp wire cap over the exposed wires *(above)*. Use the crimping jaws of the wire strippers to crimp the cap securely *(page 119)*. Repeat with the on/off switch leads and the other power cord wire. Snap on the pedal, checking that the cord reel spring fits into its slot on the pedal. Put back the control panel and the motor cover *(page 96)*.

SERVICING THE ON/OFF SWITCH (Upright type)

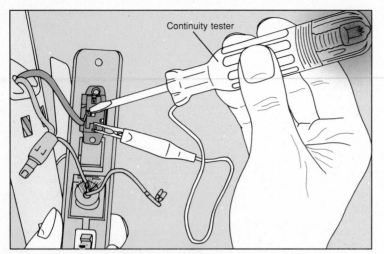

Continuity tester

1 **Testing the switch.** Unplug the vacuum cleaner and remove the switch cover plate from the handle *(page 96)*. Test a multi-speed switch as described in Tools & Techniques *(page 117)*. To test the more common single-speed switch, use a continuity tester *(page 113)*. Disconnect one wire and attach the alligator clip of the tester to the free terminal. Touch the probe of the tester to the other terminal *(above)* and flip the switch to the ON position. If the tester lights only when ON, the switch is OK. Reconnect the wire, reinstall the cover plate and test the wiring *(page 103)*. If the tester fails to light, replace the switch *(step 2)*.

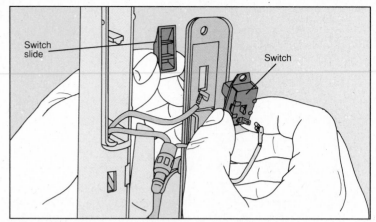

Switch slide

Switch

2 **Replacing the switch.** Use your fingers to pry off the switch slide. Buy an exact replacement switch from a vacuum cleaner service center or the manufacturer. Flick the new switch to OFF and turn it so that the lever points down. (If the on/off positions aren't marked, use a continuity tester to determine them; the tester will light with the switch in the ON position.) Fit the switch into the cover plate and snap on the switch slide *(above)*. Reconnect the wires to the switch terminals and reinstall the cover plate on the handle, making sure that all wires are tucked into the plastic housing away from the screw hole and the metal handle. Screw on the cover plate.

SERVICING THE ON/OFF SWITCH (Canister type)

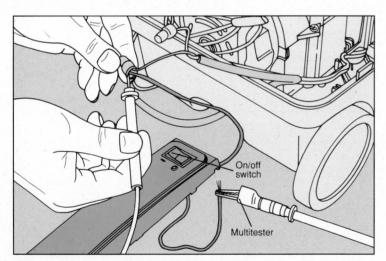

On/off switch

Multitester

1 **Testing the switch.** Unplug the vacuum cleaner. Remove the motor cover and the control panel *(page 96)*. To free a switch's wire lead for testing, trace the lead to its connection in a crimp wire cap, label the wires and use wire cutters to cut off the wire cap. Strip about 1/2 inch of insulation off the switch lead *(page 118)*. Set a multitester to test continuity *(page 113)* and attach the alligator clip of one probe to the switch lead. Trace the second switch lead to its nearest wire connection inside a crimp wire cap. Insert the second multitester probe into the cap to contact the bare wire of the switch lead *(above)*. Flip the on/off switch to the ON position. The multitester should show continuity only when ON. If the switch tests faulty, replace it *(step 2)*. If it tests OK, reconnect the cut wires with a new crimp wire cap *(page 119)*. Reinstall the control panel and the motor cover *(page 96)*.

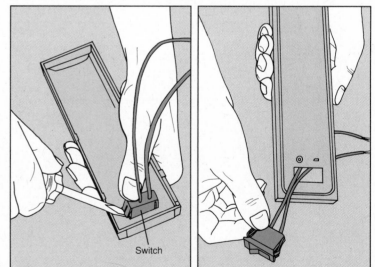

Switch

2 **Replacing the switch.** Label the wires connected with the second switch lead and cut off the crimp wire cap. Turn over the control panel. Use a screwdriver to depress the pressure tabs on one side of the switch while pushing on the corner with your thumb, until that edge of the switch slides out of the control panel. Disengage the tabs on the opposite side *(above, left)* and the switch will pop free. Buy an exact replacement from a vacuum cleaner service center or the manufacturer. To install the new switch, thread its leads through the opening in the panel *(above, right)* and snap in the switch. Reconnect both leads to their wire connections with crimp wire caps *(page 119)*. Replace the control panel and the motor cover *(page 96)*.

SERVICING THE DIRT FAN (Upright type)

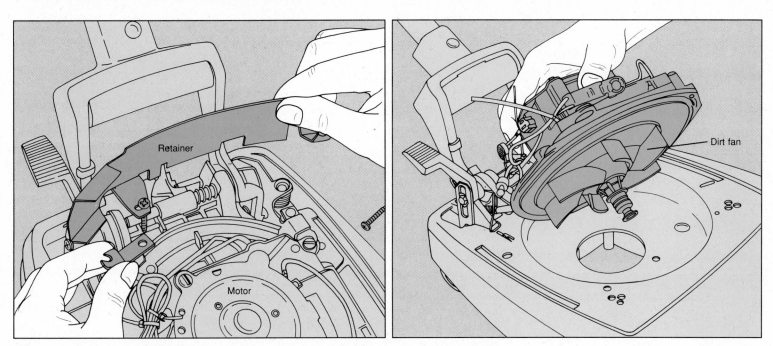

1 **Removing the fan and motor assembly.** Unplug the vacuum cleaner. Take off the bottom plate, unhook the drive belt from the motor pulley and remove the hood *(page 96)*. Use a screwdriver to unfasten the retainer and clip *(above, left)*. Locate the remaining motor mounting screws and unscrew them. Grasp the top of the motor housing and lift the fan-and-motor assembly out of the base *(above, right)*.

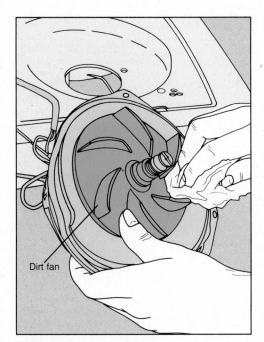

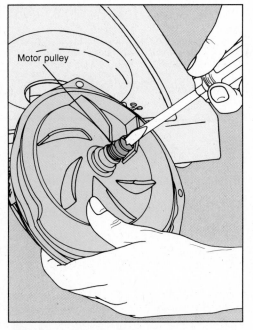

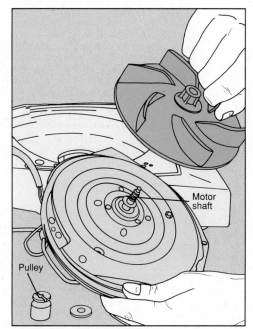

2 **Cleaning the dirt fan.** Use a soft, slightly damp cloth to wipe the fan blades *(above)* and the cavity in the vacuum cleaner base where the fan sits. Inspect a metal fan for bent blades, or a plastic fan for chipped or worn blades. Remove the fan if it is damaged *(step 3)* or to check for foreign objects caught beneath it.

3 **Removing the dirt fan.** Hold the dirt fan steady and unscrew the motor pulley *(above)*. Turn it counterclockwise; if it won't loosen, try turning it clockwise. Remove the motor pulley and its washer and set them aside. Pull the fan off the motor shaft. Note that another washer and a metal spacer are on the shaft itself; leave them in place.

4 **Replacing the dirt fan.** Purchase an exact replacement fan from a vacuum cleaner service center or the manufacturer. Slip the new fan onto the motor shaft *(above)*. Then reinstall the washer and motor pulley and screw on the pulley. Reinstall the fan-and-motor assembly, then the retainer. Put back the hood, reconnect the drive belt and reinstall the bottom plate *(page 96)*.

SERVICING THE FAN (Canister type)

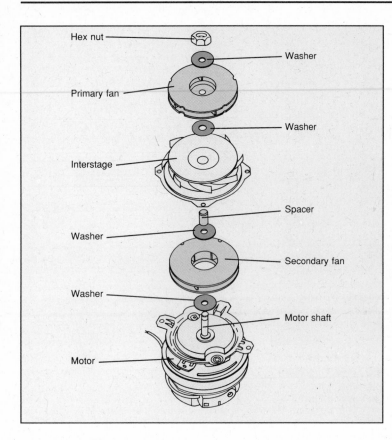

Hex nut

Washer

Primary fan

Washer

Interstage

Spacer

Washer

Secondary fan

Washer

Motor shaft

Motor

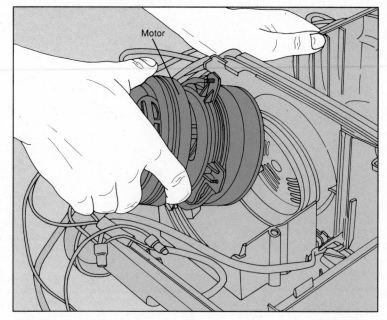

Motor

1 **Removing the canister motor.** Unplug the vacuum cleaner. Remove the bag housing lid and the dust bag. Next, remove the motor cover, the control panel, the muffler (if it has one) and the motor retainer *(page 96)*. Lift the motor out of the motor chamber *(above)* and lay it carefully on the ground.

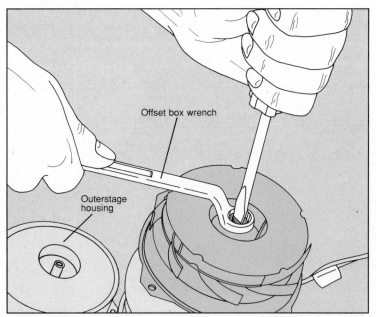

Offset box wrench

Outerstage housing

2 **Getting to the fans.** Remove the screws that hold the outerstage housing to the motor. Pull off the outerstage housing. Loosen the hex nut on the motor shaft by slipping an offset box wrench over the nut and fitting a large screwdriver into the slot of the motor shaft. (Some motors require a hex wrench rather than a screwdriver.) Hold the shaft stationary with the screwdriver and turn the wrench clockwise to loosen the nut *(above)*. If the nut won't budge, trying turning it counterclockwise. If it still sticks, squirt a few drops of penetrating oil on the shaft and let it soak for half an hour, then loosen the nut. Remove the nut and the washer.

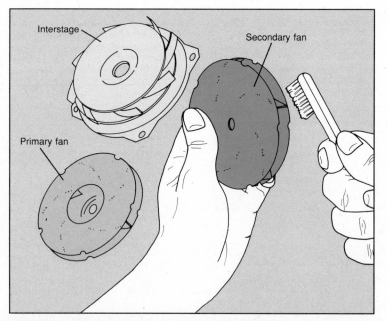

Interstage

Secondary fan

Primary fan

3 **Inspecting, cleaning and replacing the fans.** Take off the primary fan, the interstage and the secondary fan nested within the interstage. Keep track of where the spacers and washers fit between the fans. Use an old toothbrush *(above)* or a slightly damp cloth to clean the fans. Inspect them for dents or wear; the primary fan is more prone to damage since it is first in line. If a fan is damaged, take it to a vacuum cleaner service center and buy an exact replacement. Reassemble the fans using the anatomy picture *(top left)* as a guide. Reinstall the fan-and-motor assembly in the canister cavity, then reinstall the motor retainer, the muffler, the control panel and the motor cover *(page 96)*.

SERVICING THE INTERNAL WIRING

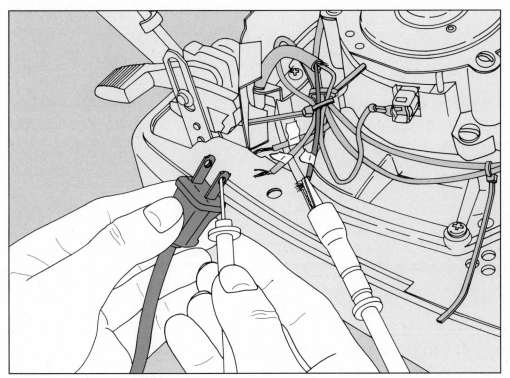

Testing the vacuum cleaner wiring. Unplug the vacuum cleaner. Access the vacuum cleaner motor by removing the bottom plate and the hood on an upright model, or the motor cover and control panel on a canister model *(page 96)*. Locate the power cord wires and disconnect one of them where it meets other wires in a crimp wire cap: Label the wires and cut off the cap with wire cutters. Flip the on/off switch to the ON position. Use a continuity tester, or set a multitester to test continuity *(page 113)*. Attach the alligator clip of one probe to the disconnected power cord wire and touch the other probe to each plug prong, in turn *(left)*. The tester should register continuity once—and only once. If the wiring tests OK, repeat the test with the other power cord wire and each plug prong. If the wiring fails the tests, suspect a loose or broken wire connection and repair it if it is accessible *(page 118)*. If you cannot locate the problem in an upright vacuum cleaner, the damaged wiring may be in the handle; take the vacuum cleaner for professional service. If the wiring tests OK, recrimp the wires *(page 119)* and check the motor brushes *(below)*.

SERVICING THE MOTOR

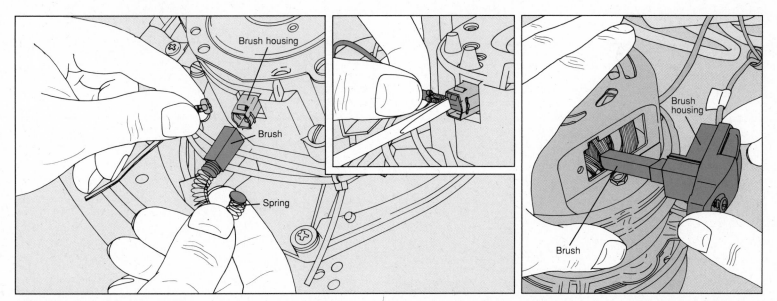

1 **Servicing the motor brushes.** Unplug the vacuum cleaner. Access the motor in your model *(page 96)*. If you suspect a binding or worn bearing on an upright model, go to step 3 to inspect the upper motor bearing. To remove a motor brush, disconnect its wire by depressing the terminal tab with a screwdriver *(inset)*. The brush will usually pop out in your hand *(above, left)*. Take out the second brush the same way. If the spring breaks off with the brush still in the housing, remove the upper motor bearing and cooling fan *(step 3)* and the upper motor housing *(step 6)*, then push out the brush from inside the housing with a small screwdriver. Some canister vacuum-cleaner motors have a brush housing that must be removed to check or change the brushes. Unscrew

the housing from the motor and pull it out *(above, right)*. Examine the brushes for pitting or wear; replace both if either is damaged. Purchase exact replacements from a vacuum cleaner service center. Fit each new brush into its housing and lock it in place with the wire terminal or by resecuring the brush housing to the motor housing. Reassemble the vacuum cleaner, reversing the steps you took to disassemble it. If the brushes are OK and the motor is a sealed unit (generally the case in canister vacuum-cleaner motors) take the vacuum cleaner for professional service. If the brushes are OK and the motor is a serviceable one, such as those found in upright vacuum cleaners, reinstall the brushes and go to step 2 to test the motor field coil.

SERVICING THE MOTOR (continued)

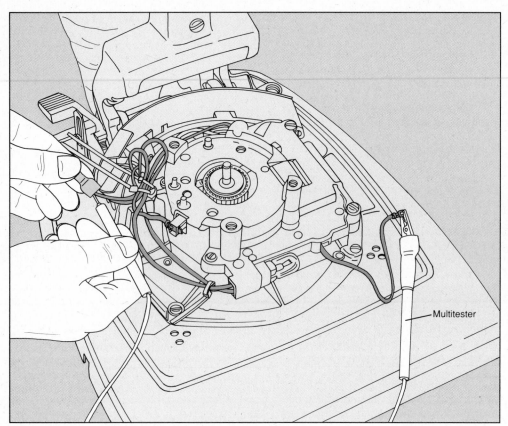

Multitester

2 **Testing the motor field coil.** To test a multi-speed motor, go to page 125. To test the field coil of the more common single-speed vacuum cleaner motor, disconnect one of the motor's wire leads. If the lead is clipped to a motor brush, release it *(step 1)*. If the lead is connected to other wires in a crimp wire cap, label the wires and cut off the cap with wire cutters. Set a multitester to RX100 *(page 113)* and attach the alligator clip of one probe to the end of the disconnected lead. Touch the other probe to the end of the second motor lead. If it is still connected to other wires in a crimp wire cap, try slipping the narrow tip of the probe into the cap, making sure it is in contact with the lead's copper filaments *(left)*. If there isn't enough space in the cap for the probe, label the wires, cut off the cap and touch the probe to the motor lead. The multitester should register partial resistance. If the field coil has continuity or infinite resistance, it is faulty; take the vacuum cleaner for professional service. If it tests OK, remove the upper motor bearing and fan *(step 3)* and test the commutator *(step 4)*.

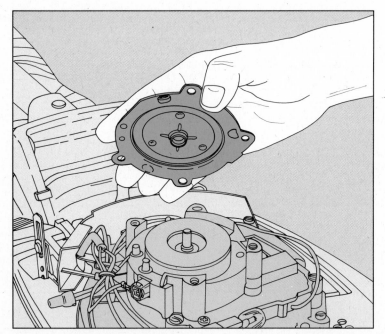

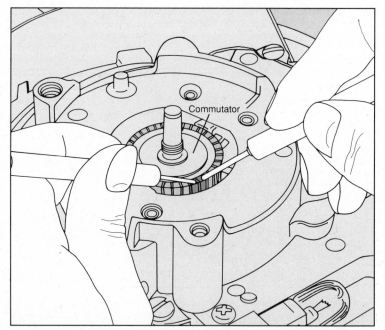

Commutator

3 **Replacing the motor bearing.** The upper motor bearing consists of a sleeve inserted in a metal shield; to remove it from the upper motor casing, unscrew its mounting screws and lift it off *(above)*. Examine the bearing; if worn, install an exact replacement purchased at a vacuum cleaner service center. To reach the commutator for testing, remove the cooling fan: Hold the motor pulley underneath with one hand, while you turn the fan with the other. If it doesn't come off when turned counterclockwise, try turning it clockwise.

4 **Performing a bar-to-bar commutator test.** To test whether there is a break in the armature wiring, set a multitester to RX1 *(page 113)*. Place the probes on adjacent commutator bars *(above)*; the multitester should indicate low resistance. Repeat this test between all adjacent bars on the commutator, and look for similar low resistance each time. If the commutator fails this test, remove the rotor *(step 6)* for service. If the commutator tests OK, go to step 5 to test for a ground fault in the motor.

SERVICING THE MOTOR (continued)

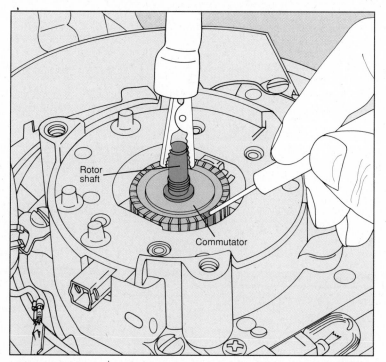

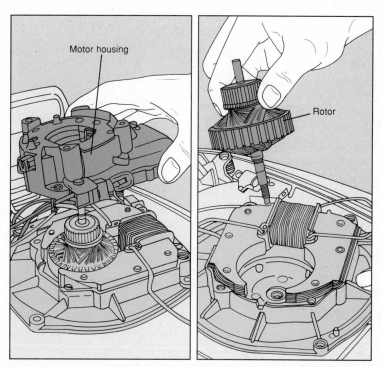

5 **Testing for a ground fault.** Set a multitester to test continuity *(page 113)*. Attach a multitester probe with an alligator clip to the rotor shaft, and touch the second probe to each commutator bar, in turn *(above, right)*. The multitester should show no continuity. If the armature wiring fails this test, replace the rotor *(step 6)*. If it tests OK, remove the rotor and clean the commutator.

6 **Removing the rotor.** Remove the brushes *(step 2)*. Unscrew the upper motor housing from the lower housing, lift it off *(above, left)* and put it aside. Remove the dirt fan from underneath *(page 101)*. Take hold of the motor shaft, pull out the rotor *(above, right)*, and inspect it for service or replacement *(step 7)*.

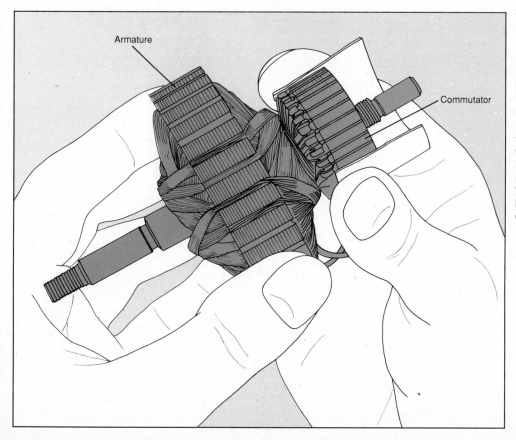

7 **Cleaning the commutator.** Check the commutator for pitting or carbon buildup. To remove roughness, smooth the surface with a piece of fine emery cloth. Use an old toothbrush to clean out grit and dirt between the bars. If the armature wiring is broken or shorted, or if the commutator bars are damaged, replace the rotor. Buy an exact replacement at a vacuum cleaner service center or from the manufacturer. Position it carefully inside the lower motor casing, resecure the dirt fan, and reinstall the fan-and-motor assembly *(page 101)*. Reinstall the upper motor housing and the motor brushes before putting back the hood, drive belt and bottom cover *(page 96)*.

SEWING MACHINES

Through the interaction of drive shafts, gears, pulleys and belts, a sewing machine harnesses the power of a universal motor to perform the delicate task of bringing two threads together, producing as many as 1,200 uniform stitches a minute. A typical sewing machine is illustrated on page 107. On some models the motor is hidden inside the housing. The motor on this model is mounted on the exterior. If the motor, foot pedal, power cord or on/off switch are faulty on the model shown, they all must be replaced as a unit. Other differences are found in the tension and bobbin assemblies and in the change from manual controls to electronic controls.

In spite of the variation among brands, all sewing machines use similar parts to perform the same function, that of looping the upper, or needle, thread around the lower, or bobbin, thread in a lock stitch. As a stitch is formed, the feed dog moves the fabric to ensure that the stitches are made in succession rather than on top of each other.

Given the sewing machine's reliance on the precise balance and timing of internal parts, many repairs are best left to the experience of a professional. But fortunately for the home sewer, the sewing machine is by far the sturdiest of small appliances. In fact, it is not uncommon to find sewing machines that are still working after thirty years. When sewing machine problems do arise, it is often because the user omitted an essential step, such as not tightening the hand wheel knob enough to reengage the stitching mechanism after winding the bobbin. Use the Troubleshooting Guide *(right)* as a checklist of procedures to try, and consult your owner's manual before assuming a mechanical or electrical fault in the machine.

To avoid problems always use the appropriate needle and thread for the fabric, matching the size and type of bobbin thread to the needle thread. When sewing, start up slowly and don't pull the fabric; let it feed naturally. If a problem does arise, simply cleaning and lubricating the machine, adjusting exterior tension or pressure regulators or replacing parts around the needle and bobbin case may be all that is needed. Specific instructions for maintenance and adjustment of your particular model are contained in the owner's manual. If you don't have one, order it from the manufacturer or an authorized service center before attempting repair.

TROUBLESHOOTING GUIDE

SYMPTOM	PROCEDURE
Motor does not run at all	Plug in and turn on machine; reset breaker or replace fuse *(p. 112)* □○
	Test and replace power cord/foot pedal assembly *(p. 109)* ◖○▲ or replace motor assembly *(p. 109)* ◖○
Motor runs but sewing machine doesn't work	Tighten hand wheel knob and disengage bobbin winder
	Adjust, replace drive belt *(p. 109)* ◖○
	Clean, lubricate machine *(p. 107)* □◕
	Replace motor assembly *(p. 109)* □○
Sewing machine is noisy	Replace needle if bent or damaged; consult owner's manual
	Clean, lubricate machine *(p. 107)* □◕
	Loosen drive belt *(p. 109)* □○
Sewing machine runs slowly	Tighten hand wheel knob and disengage bobbin winder; consult owner's manual
	Adjust, replace drive belt *(p. 109)* ◖○
	Clean, lubricate machine *(p. 107)* □◕
Sewing machine jams	Remove pin or other foreign object from bobbin case; clean and lubricate machine *(p. 107)* □◕
Thread breaks repeatedly	Use correct needle and thread for type of fabric and install correctly; replace bent needle; consult owner's manual
	Service thread path *(p. 108)* □○
	Loosen thread tension by adjusting tension dial
	Clean feed dog, bobbin case and tension discs *(p. 107, step 1)* □○
	Adjust bobbin-winder tension disc *(p. 108)* □○; rewind bobbin evenly
Bobbin doesn't wind	Replace bobbin-winder friction ring *(p. 108)* □○; consult owner's manual
Bobbin winds unevenly	Adjust bobbin-winder tension disc *(p. 108)* □○
Fabric doesn't feed properly or is damaged while stitching	Replace bent presser foot or needle; don't pull fabric, let it feed naturally; consult owner's manual
	Clean feed dog and bobbin case *(p. 107, step 1)* □○
	Service thread path *(p. 108)* □○
Needle breaks	Use correct needle and thread for type of fabric; use correct needle plate for type of stitch; consult owner's manual
Stitches irregular	Adjust bobbin-winder tension disc *(p. 108)* □○ and rewind bobbin evenly; consult owner's manual
	Service thread path *(p. 108)* □○
	Clean bobbin case and tension discs *(p. 107, step 1)* □○

DEGREE OF DIFFICULTY: □ Easy ◖ Moderate ■ Complex
ESTIMATED TIME: ○ Less than 1 hour ◕ 1 to 3 hours
▲ Special tool required

SEWING MACHINES

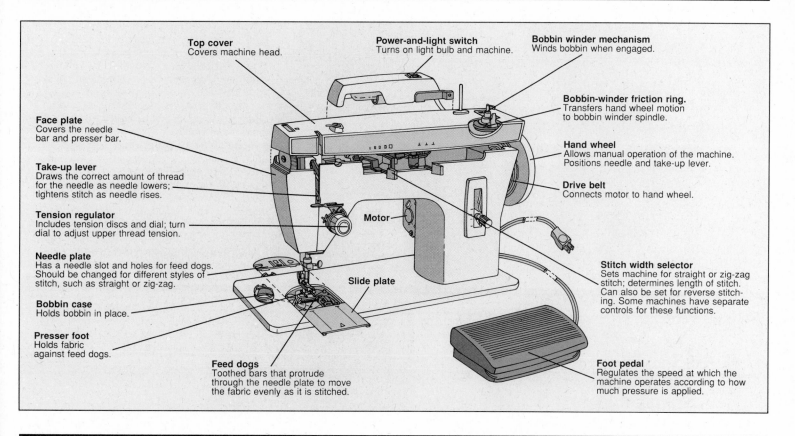

Top cover
Covers machine head.

Power-and-light switch
Turns on light bulb and machine.

Bobbin winder mechanism
Winds bobbin when engaged.

Face plate
Covers the needle bar and presser bar.

Take-up lever
Draws the correct amount of thread for the needle as needle lowers; tightens stitch as needle rises.

Tension regulator
Includes tension discs and dial; turn dial to adjust upper thread tension.

Needle plate
Has a needle slot and holes for feed dogs. Should be changed for different styles of stitch, such as straight or zig-zag.

Bobbin case
Holds bobbin in place.

Presser foot
Holds fabric against feed dogs.

Feed dogs
Toothed bars that protrude through the needle plate to move the fabric evenly as it is stitched.

Motor

Slide plate

Bobbin-winder friction ring.
Transfers hand wheel motion to bobbin winder spindle.

Hand wheel
Allows manual operation of the machine. Positions needle and take-up lever.

Drive belt
Connects motor to hand wheel.

Stitch width selector
Sets machine for straight or zig-zag stitch; determines length of stitch. Can also be set for reverse stitching. Some machines have separate controls for these functions.

Foot pedal
Regulates the speed at which the machine operates according to how much pressure is applied.

CLEANING AND LUBRICATING

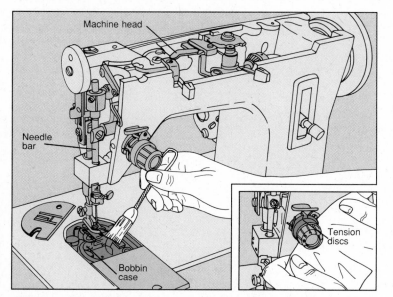

Machine head

Needle bar

Bobbin case

Tension discs

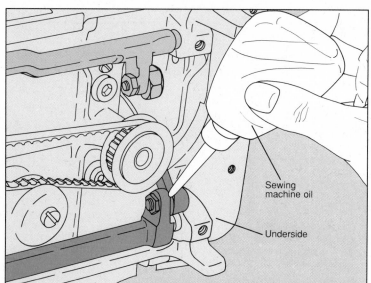

Sewing machine oil

Underside

1 Cleaning. Unplug the sewing machine and remove the thread. Unscrew and remove the top cover and face plate. Use a small, stiff brush to wipe dirt and lint off the machine head and the needle bar area. Raise the presser foot and the needle to their highest position, open the slide plate and remove the bobbin and needle plate. Brush away dirt and lint from the feed dogs and bobbin case *(above)*, and remove any tangled thread with your fingers. Use a soft cloth to clean between the tension discs on the tension regulator *(inset)*. To clean underneath, remove the carrying case base or, if the machine is in a cabinet, tilt it back.

2 Lubricating. Consult your owner's manual for the exact spots to oil inside the machine head, needle bar area, bobbin case and underside. Only use sewing machine oil recommended by the manufacturer. On the model shown here, put one drop of oil anywhere that two metal parts rub together *(above)*; locate these spots by looking inside the machine as you turn the hand wheel toward the front of the machine. After oiling, plug in the machine and run it slowly for a few minutes. Then reassemble the machine, rethread it, and stitch a scrap of cotton to absorb any excess oil.

SERVICING THE THREAD PATHS

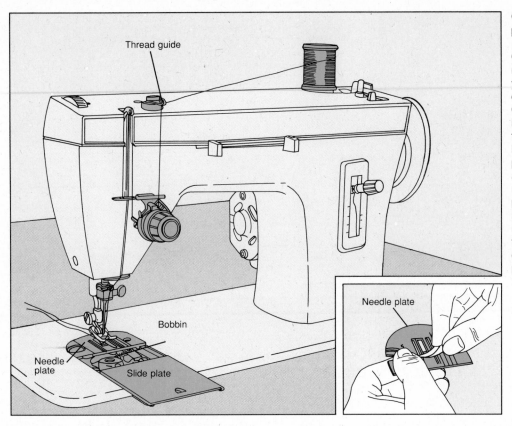

Clearing the thread path. Follow the thread path from the first thread guide to the needle *(left)*, checking for tangled or knotted thread. Check your owner's manual to make sure the machine is correctly threaded. Then, with your finger, feel for burrs, nicks or roughness on any metal parts that contact the thread. Open the slide plate, remove the bobbin and inspect the bobbin case and hook. Remove the needle plate and examine the needle hole. Sand any rough areas with fine emery paper *(inset)* and blow away loose grit. Reassemble the machine. If burrs or nicks remain on the needle, presser foot, needle plate or bobbin case, replace them with exact replacement parts purchased from an authorized service center; if they remain on the thread guides, take-up lever, bobbin hook or tension regulator, take the machine for professional service.

SERVICING THE BOBBIN WINDER

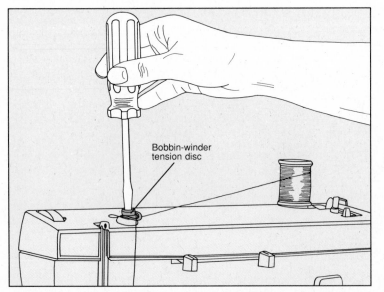

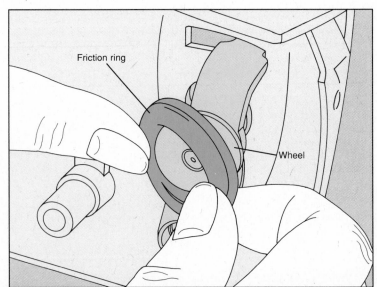

Adjusting the bobbin-winder tension disc. With the machine set to wind the bobbin, locate the screw holding the bobbin-winder tension disc, on top of the machine. Insert a screwdriver in the slot of its screw, and press on the foot pedal to begin winding the bobbin. Adjust the tension by turning the screw counterclockwise to loosen or clockwise to tighten *(above)*, while watching the bobbin as it winds. Adjust until the thread winds evenly from top to bottom on the bobbin. On a correctly wound bobbin, the thread is wound uniformly, never in a cone shape.

Replacing the bobbin winder ring. Unplug the machine. Unscrew and remove the top cover. Locate the bobbin winder spindle on top of the cover, then turn over the cover to find the wheel attached to this spindle on the inside. If the rubber friction ring is worn or cracked, pull it off the wheel. Buy an exact replacement friction ring from an authorized service center or the manufacturer. Install the ring by pushing it onto the wheel so that it fits in the groove. Reinstall the top cover and tighten the screws.

SERVICING THE DRIVE BELT

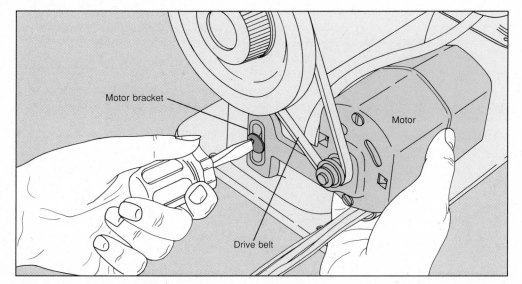

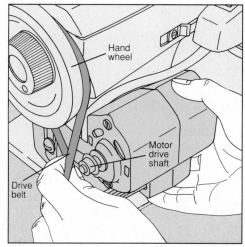

1 **Adjusting the drive belt.** Unplug the machine. Locate the motor bracket screw and loosen it just enough to allow you to move the motor bracket, while supporting the weight of the motor with your other hand *(above)*. If the machine has been running slowly, loosen the belt by raising the motor, keeping the arm of the motor bracket horizontal. If the machine has been running unevenly, tighten the belt by lowering the motor. The belt should be just tight enough so that it doesn't slip. If you have lowered the motor as far as it will go and the belt is still loose, purchase an exact replacement belt from an authorized service center or the manufacturer and install it *(step 2)*. When the belt is adjusted correctly, tighten the motor bracket screw, making sure that the indentation on the back of the bracket fits securely over the ridge on the machine.

2 **Replacing the drive belt.** Loosen the motor bracket screw while supporting the motor with your other hand and lift the belt off the motor drive-shaft pulley *(above)*. Set the motor down and remove the belt from around the hand wheel. To install the new belt, place it around the hand wheel, then lift up the motor and put the belt around its drive shaft pulley. Adjust the belt tension *(step 1)* and tighten the motor bracket screw.

SERVICING THE POWER ASSEMBLY

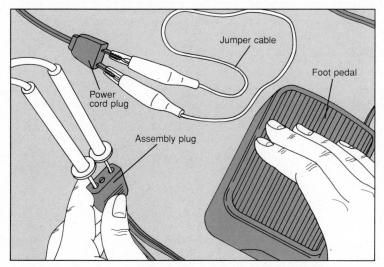

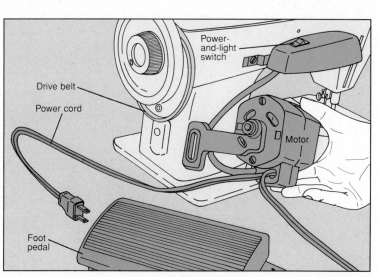

Testing and replacing the power cord/foot pedal assembly. If your model has a power cord/foot pedal assembly that plugs into the sewing machine, unplug it from the wall outlet and machine. Set a multitester to RX1 *(page 113)* and clip a jumper cable to the prongs of the power cord plug as shown. Insert the tester probes into two of the three receptacle holes in the assembly plug and depress the foot pedal completely. There should be continuity. Repeat the test with the probes in the other two possible pairing of holes. If the assembly tests OK, take the machine for service. If there is no continuity for any of the tests, buy an exact replacement power cord/foot pedal assembly from an authorized service center.

Replacing the motor, power-and-light switch, foot pedal and power cord assembly. On some models you must have the motor, power-and-light switch, foot pedal and power cord serviced or replaced as a single unit. Remove the motor first *(above)*, supporting it in one hand as you remove the motor bracket screw with the other. Keep track of the washer. Take the drive belt off the motor's drive shaft, and set the motor down. Unscrew and remove the bracket securing the light to the machine. Take the assembly to an authorized service center and have it serviced or buy an exact replacement. To install the assembly, reverse the steps taken to remove it, and adjust the tension of the drive belt *(step 1, above)* before turning on the machine.

TOOLS & TECHNIQUES

This section introduces basic tests and procedures common to almost all small appliance repairs, from testing switches and motors to repairing broken wire connections and replacing power cords. Testing and replacing rechargeable nicad batteries is included on page 122. Servicing universal and shaded pole motors is described on pages 123-125. Most of the tools used in this book may be found in an all-purpose tool kit, but visit an electronics supplies store to purchase specialized items such as the miniature butane torch for silver soldering, and the parts required to make the capacitor discharging tool.

When a small appliance stops working, determine whether the problem originates outside the unit before you take the appliance apart. Calculate the electrical load on the circuit *(page 114)* and move several appliances to another circuit if necessary. Examine the appliance's power cord *(page 116)*; a frayed cord or damaged plug are among the more common causes of small appliance failure.

Getting inside a small appliance to make a repair is often the most difficult part of the job. Tips on disassembling your appliance appear on page 115. Check for hidden screws and tabs rather than forcing panels or casings apart.

The electrical tests described in this book require a multi-tester or a continuity tester. Both are battery powered to send a small electrical current through the part being tested. Before conducting any test, refer to the section in this chapter on troubleshooting with electrical testers *(page 113)*.

Ensure the best repair results by using the right tools for the job. To avoid stripping screw heads, choose a screwdriver that

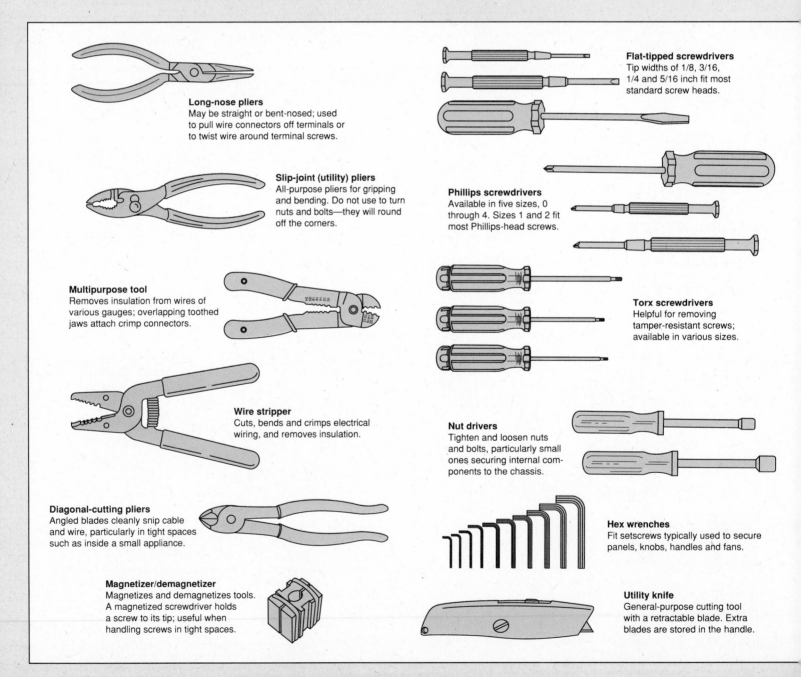

Long-nose pliers
May be straight or bent-nosed; used to pull wire connectors off terminals or to twist wire around terminal screws.

Slip-joint (utility) pliers
All-purpose pliers for gripping and bending. Do not use to turn nuts and bolts—they will round off the corners.

Multipurpose tool
Removes insulation from wires of various gauges; overlapping toothed jaws attach crimp connectors.

Wire stripper
Cuts, bends and crimps electrical wiring, and removes insulation.

Diagonal-cutting pliers
Angled blades cleanly snip cable and wire, particularly in tight spaces such as inside a small appliance.

Magnetizer/demagnetizer
Magnetizes and demagnetizes tools. A magnetized screwdriver holds a screw to its tip; useful when handling screws in tight spaces.

Flat-tipped screwdrivers
Tip widths of 1/8, 3/16, 1/4 and 5/16 inch fit most standard screw heads.

Phillips screwdrivers
Available in five sizes, 0 through 4. Sizes 1 and 2 fit most Phillips-head screws.

Torx screwdrivers
Helpful for removing tamper-resistant screws; available in various sizes.

Nut drivers
Tighten and loosen nuts and bolts, particularly small ones securing internal components to the chassis.

Hex wrenches
Fit setscrews typically used to secure panels, knobs, handles and fans.

Utility knife
General-purpose cutting tool with a retractable blade. Extra blades are stored in the handle.

fits, and don't use excessive force. Magnetize screwdrivers to avoid time-consuming hunts for a screw dropped inside a unit. Clean metal tools with a few drops of light machine oil. To remove rust, rub with fine steel wool or emery paper. Keep tools in a sturdy box, with a lock if stored around children.

To operate efficiently, electrical appliances rely on a smooth flow of current through the unit's circuitry. Often, simply inspecting wire connections can uncover the source of a problem. Damaged wire ends can be stripped of their insulation, then reconnected as described in this chapter. Always buy identical replacement wire; if in doubt about the type or gauge required, bring the old wire with you.

Refer to the appliance owner's manual before undertaking any repair; even changing a power cord may void the warranty provided by the manufacturer. Always disconnect an appliance before servicing it and keep the power cord plug in plain view on your work surface. Make sure that the work area is free of clutter and well lit. Work patiently and methodically; never take short cuts. Label all wires for correct reconnection. Above all, keep in mind that even seemingly complicated tasks are seldom more than a sequence of easy steps.

Always substitute a faulty component with an identical replacement. To ensure that no wires are pinched or disconnected, perform a cold check for leaking voltage *(page 114)* after reassembling an appliance but before plugging it back in. Consider purchasing the ground-fault circuit interrupter described in this chapter; it immediately cuts off power to an appliance should leaking current be detected.

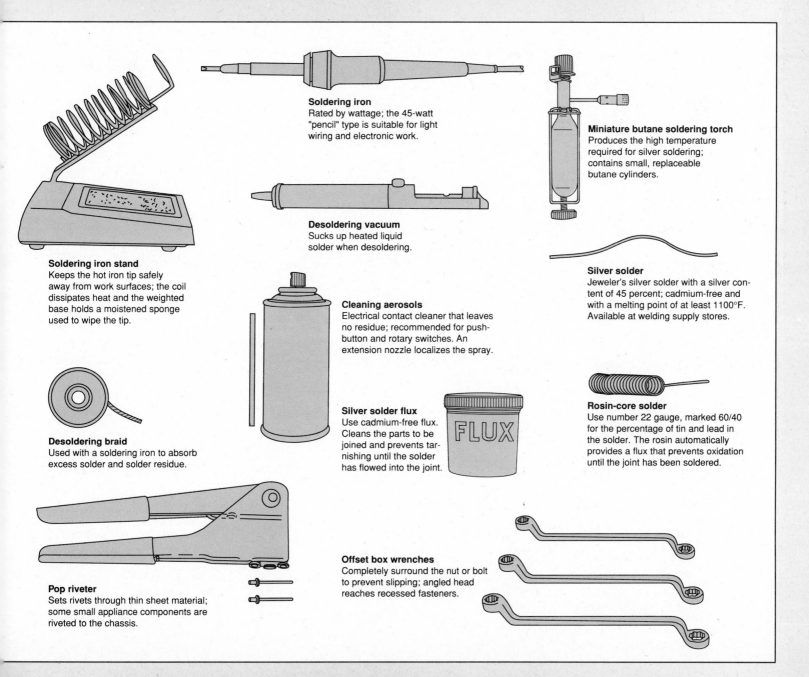

Soldering iron
Rated by wattage; the 45-watt "pencil" type is suitable for light wiring and electronic work.

Miniature butane soldering torch
Produces the high temperature required for silver soldering; contains small, replaceable butane cylinders.

Desoldering vacuum
Sucks up heated liquid solder when desoldering.

Silver solder
Jeweler's silver solder with a silver content of 45 percent; cadmium-free and with a melting point of at least 1100°F. Available at welding supply stores.

Soldering iron stand
Keeps the hot iron tip safely away from work surfaces; the coil dissipates heat and the weighted base holds a moistened sponge used to wipe the tip.

Cleaning aerosols
Electrical contact cleaner that leaves no residue; recommended for push-button and rotary switches. An extension nozzle localizes the spray.

Silver solder flux
Use cadmium-free flux. Cleans the parts to be joined and prevents tarnishing until the solder has flowed into the joint.

FLUX

Rosin-core solder
Use number 22 gauge, marked 60/40 for the percentage of tin and lead in the solder. The rosin automatically provides a flux that prevents oxidation until the joint has been soldered.

Desoldering braid
Used with a soldering iron to absorb excess solder and solder residue.

Offset box wrenches
Completely surround the nut or bolt to prevent slipping; angled head reaches recessed fasteners.

Pop riveter
Sets rivets through thin sheet material; some small appliance components are riveted to the chassis.

CHECKING THE MAIN SERVICE PANEL

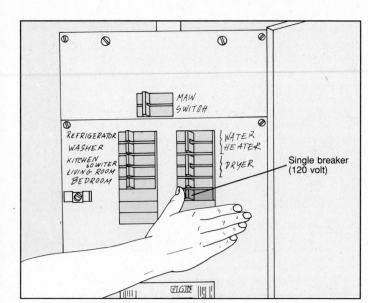

Resetting a tripped circuit breaker. When an electrical circuit is overloaded, the circuit breaker toggle automatically flips to OFF or an intermediate position, shutting off power. When this happens, turn off or unplug the appliances on the circuit. Then reset the circuit breaker by pushing it fully to the OFF position and then back to ON *(above)*. If the circuit breaker trips again, inspect the appliances, especially their power cords *(page 116)*, for a short circuit and correct the problem, or have the appliance or circuit serviced professionally.

Replacing a fuse. Electrical circuits in older systems are protected by plug-type fuses. When a fuse blows, turn off or unplug the appliances on the circuit. Then unscrew the fuse and replace it with one of identical amperage *(above)*. A complete break in the fuse's metal strip indicates a circuit overload; move one or two appliances to another circuit. A discolored fuse points to a short circuit; inspect the appliances, especially their power cords *(page 116)*. If you can't correct the problem, have the appliance or circuit serviced professionally.

CALCULATING ELECTRICAL LOAD

Determining a circuit overload. If you suspect that a circuit is overloaded, calculate the existing load and compare it to the capacity of the circuit, which is indicated by the amperage rating of the fuse or circuit breaker. To calculate the existing load on a circuit, list all the appliances and fixtures on the circuit, along with the wattage rating for each device. On a small appliance, this information is usually stamped on a plate *(below)* or on its housing. On a lighting fixture, it is usually near the socket. Typical wattage ratings of many small appliances are listed in the chart at right. Add the wattage ratings for all appliances and fixtures on the circuit, then divide by 120 volts to convert to amperes. If the total is higher than the amperage of the circuit, the circuit is overloaded. Move a high-wattage appliance such as a hair dryer or toaster oven to another circuit, or have an electrician run a new circuit from the service panel.

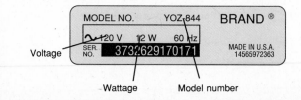

TYPICAL SMALL-APPLIANCE LOADS

APPLIANCE	APPROXIMATE WATTAGE RATING	APPLIANCE	APPROXIMATE WATTAGE RATING
Blender	200-400	Kettle	1200-1400
Can opener	150	Microwave oven	500-800
Coffee maker	600-1000	Mixer	100-225
Food processor	500	Sewing machine	100
Frying pan	1000-1200	Shaver	15
Hair dryer	400-1500	Toaster	800-1200
Humidifier	50-150	Toaster oven	1500
Iron	1200	Vacuum cleaner	300-600

GETTING HELP WHEN YOU NEED IT

Keeping small appliances in working order. Many small appliance failures can be prevented by careful maintenance and correct usage. Follow the use and care tips in this book and review your owner's manual. If you don't have a manual, locate the appliance's model number (stamped on a plate attached to the housing or on the housing itself) and use it to order a manual from the manufacturer.

Clean appliances regularly and check the power cord periodically for signs of wear and tear. Pay attention to the sound, smell and feel of a smoothly operating appliance, so that you know immediately when it begins to show strain. For example, an unusually high-pitched drone from a motorized appliance is an indication that the motor has to work too hard to overcome a fault in related parts, such as a fan or gears. Catching the problem early may prevent damage.

Finding what you need to make effective repairs. Repairing the appliance may call for replacing a broken part. Armed with the model number and the name of the part, contact the supplier to make sure it is in stock. Finding a supplier for your appliance may require several phone calls. You have a variety of sources from which to choose:

Small appliance parts dealers. If the store does not carry a specific part, the dealer may be able to order the part or tell you where to find it.

Small appliance repair shops. Often these shops are the only source of discontinued and hard-to-find replacement parts.

Authorized service centers. Some service centers specialize in repairing one or two types of appliances only.

Hardware stores. Many stock tools, wiring and other electrical supplies, but their salespeople may not be specialized.

Manufacturers. Get the address and telephone number of the manufacturer from the owner's manual or by asking a parts dealer or the appliance retailer.

If the appliance requires professional service, choose carefully, and ask friends for references. Your local Better Business Bureau or consumer rights office may help you select a reputable technician, who should be trained to service your brand of appliance.

You and your warranty. Before attempting any repair, check the warranty provided by the manufacturer. If the warranty is still in effect, you will void it if you undertake the repair yourself; take the appliance to an authorized service center.

TROUBLESHOOTING WITH ELECTRICAL TESTERS

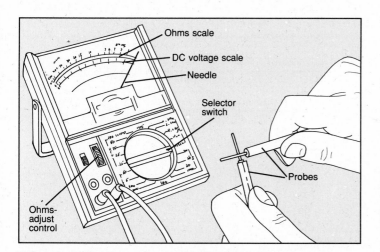

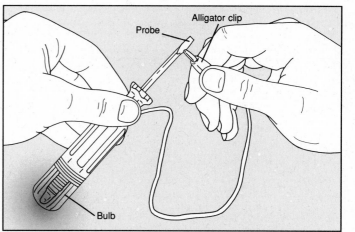

Using a multitester and a continuity tester. A multitester can indicate whether an electrical circuit is completed (continuity); it can measure, in ohms, the actual amount of resistance in a circuit (resistance); and it can measure, in volts, the strength of current passing through the circuit (voltage). Before each test for continuity or resistance, ensure a precise reading by "zeroing" the multitester: Set the multitester to the ohms setting you plan to use, and touch the probes together *(above, left)*. The needle will sweep from left to right toward ZERO; turn the ohms-adjust control until the needle aligns directly over ZERO.

Unless otherwise specified, setting a multitester to test continuity or resistance means setting it to RX1. Touch the probes to the designated terminals or wire ends. If the circuit has continuity and no resistance, the needle will indicate zero ohms. If the circuit is incomplete, the needle will indicate infinite resis-

tance. If the circuit has some resistance (as in the case of motor windings or heating elements), but not enough to prevent current flow, the needle will move to a point between zero ohms and infinite resistance.

The multitester's capacity to measure low voltage makes it ideal for testing the charging circuit on a rechargeable appliance. Test a low voltage DC circuit by setting the multitester to 10 volts on the DCV scale.

Unlike the multitester, the continuity tester can only show whether a circuit is completed. First, check the tester battery by touching the alligator clip to the probe *(above, right)*; the bulb should light. Then, to test the circuit, touch the probes to the designated points. If the circuit has continuity, the bulb will glow. If the bulb fails to light, there is either a break or partial resistance in the circuit.

PREVENTING ELECTRICAL HAZARDS

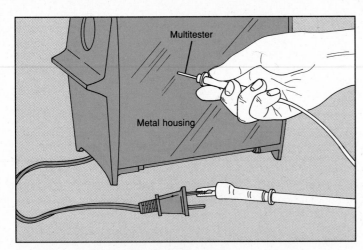

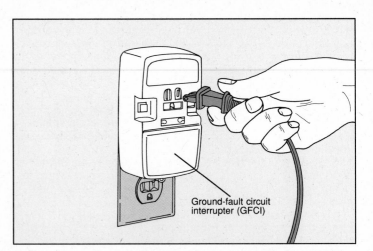

A cold check for leaking voltage. After reassembling an appliance, but before plugging it in, check that current cannot leak to the chassis. Set a multitester to RX1000 *(page 113)*. Clip one probe to a plug prong and touch the other probe to the metal housing *(above)*. Move the control knob to ON. The multitester should indicate infinite resistance. Repeat the test with the other plug prong. If the housing is plastic, touch the probe to unpainted metal parts such as mounting screws and metal trim. If the multitester indicates anything but infinite resistance, disassemble the appliance and check for wires touching the chassis. If you cannot correct the problem, take the appliance for professional service.

Using a ground-fault circuit interrupter (GFCI). This device provides protection against electrical shock. It monitors the flow of electrical current and can detect even small amounts of current leaking to ground. The moment an electrical irregularity is detected, the GFCI automatically shuts off the circuit. In homes built or wired more than fifteen years ago, it is unlikely that GFCI's are permanently installed in the main service panel. As an extra safety precaution, purchase a GFCI from a hardware store or an electrical parts supplier and plug it into a conventional receptacle according to the manufacturer's instructions *(above)*.

DISCHARGING A CAPACITOR

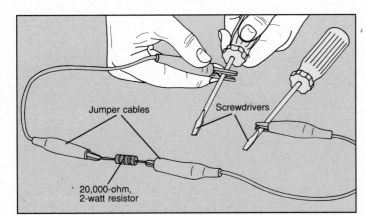

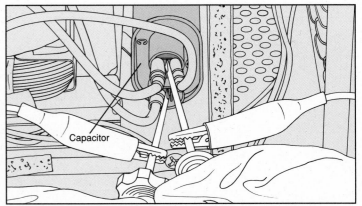

1 Making a capacitor discharging tool. This simple tool drains capacitors of stored charge without damage to them or injury to you. Assemble two jumper cables with alligator clips; a 20,000-ohm, 2-watt resistor, available at an electronics supplies store; and two screwdrivers with insulated handles. Clip a jumper cable to each resistor lead. Then clip a jumper cable to the blade of each screwdriver *(above)*. To discharge a capacitor, go to step 2.

2 Discharging a capacitor. Caution: Capacitors store potentially dangerous voltage. Unplug and disassemble the appliance *(page 115)*. Locate, but do not touch, the capacitor, a metal canister with two terminals usually mounted near the motor. Wait five minutes. Then, taking care not to touch any internal components, touch a screwdriver blade of the discharging tool to each capacitor terminal simultaneously, for one second *(above)*.

TIPS ON DISASSEMBLY AND REASSEMBLY

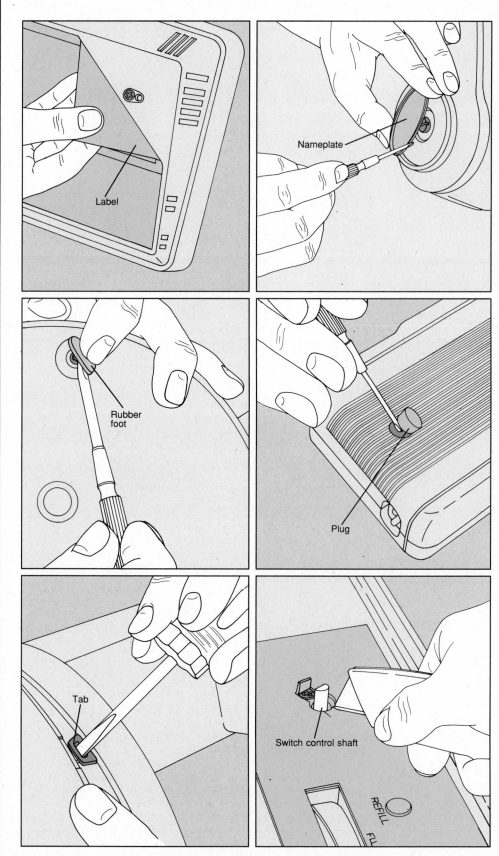

Finding concealed fasteners. Before trying to disassemble an appliance, review the owner's manual. Check whether the appliance is still under warranty. If it is, take the unit for service. If not, set up for repair on a clean, well-lit work table. Magnetize screwdrivers before you start to work by passing their heads through a magnetizer/demagnetizer *(page 110)*. Label small jars or film canisters for storing screws and small parts.

First study the appliance to determine how the housing is put together. On many appliances, the fasteners are recessed or completely hidden from view. Check for screws under labels and nameplates; run your fingers over the label to feel for screw heads and gently peel back a corner of the label *(top left)*. Undo the nameplate's mounting screws or slip a screwdriver blade under a glued plate to pry it off *(top right)*. Lift off rubber, felt or plastic feet with your fingers or the tip of a screwdriver to find screws *(center left)*. Sometimes the only sign of a hidden screw is a small circular seam in the housing, indicating a plug set flush with the surface. Work the tip of a small screwdriver gently along the seam to pry out the plug *(center right)*. To release a stubborn screw, press the screwdriver firmly into the screw head and snap any glue seal with a sharp counterclockwise twist of the tool.

If you cannot find any assembly screws, the housing may be held together with tabs. Slip a screwdriver blade into the seam of the housing and gently work the housing apart. Then press the tab with the blade tip to release it from its notch *(bottom left)*.

Disassemble only what is needed to reach the parts being repaired. As you work, write down the sequence of disassembly steps for reference in reassembly. Once a part's mounting screws have been removed, carefully lift the part out of the unit without pulling wires or dislodging other parts. If the part's mounting fasteners are not visible, do the same detective work as for housing screws. The part may be secured from the exterior: Look beneath labels and control knobs, and as a last resort, cut away part of the control panel with a sharp utility knife to expose the screw heads *(bottom right)*.

Before reinstalling the housing, make sure all parts are in place and that wires are secure. After reassembly but before plugging in the appliance, perform a cold check for leaking voltage *(page 114)*.

SERVICING A 120-VOLT POWER CORD

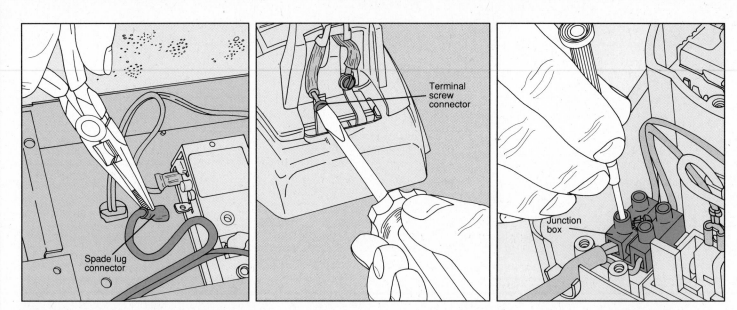

1 **Disconnecting a power cord wire.** Unplug the appliance. Consult the appliance's chapter and the tips on disassembly *(page 115)* to access the power cord terminals. If the power cord wires are connected by spade lugs, pull one off its terminal *(above, left)*; if they are secured by screw-on connectors, unscrew one *(above, center)*; if they are clamped inside a junction box by screws, remove one screw and pull the wire free *(above, right)*. If the wires are connected to other wires by a crimp wire cap, label the wires and cut off the cap; if they are soldered, desolder one wire *(page 120)*.

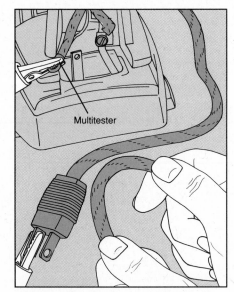

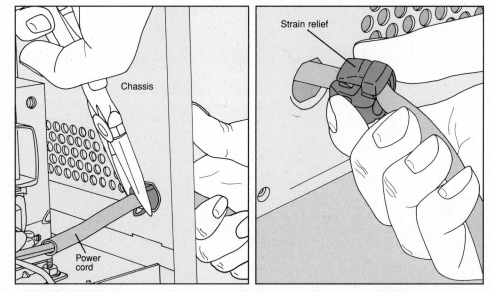

2 **Testing the power cord.** Set a multitester to test continuity *(page 113)*. Clip one probe to a plug prong and the other probe to the end of one power cord wire, then the other. The tester should show continuity only once. Bend and twist the entire cord during each test *(above)*; the needle should not move. If the power cord tests OK, reconnect it, reassemble the appliance and cold check for leaking voltage *(page 114)*. If the cord tests faulty, go to step 3 to replace it.

3 **Replacing the power cord.** Disconnect the remaining power cord wire and remove the cord. If the power cord is held to the chassis with a strain relief grommet, squeeze the grommet with pliers and push it through its hole *(above, left)*. Then pull the power cord free and remove the grommet, keeping it for the new power cord. Buy an exact replacement power cord of the correct gauge and insulation *(page 118)* from an electrical parts supplier. Cut off a short length of the outer insulation at the end of the cord with a utility knife, then strip back the inner insulation from each wire end *(page 118)*. If the old power cord had terminal connectors, install identical connectors *(page 119)*. Thread the power cord through the chassis, install the strain relief grommet on the cord *(above, right)* and push the grommet back into its hole. Clip, screw or solder *(page 120)* the wires to their terminals, or install new crimp wire caps *(page 119)*. Reassemble the appliance, reversing the steps taken to disassemble it, and cold check for leaking voltage *(page 114)*.

SERVICING SWITCHES

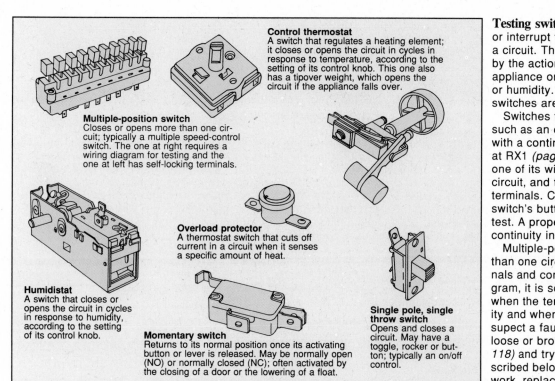

Multiple-position switch
Closes or opens more than one circuit; typically a multiple speed-control switch. The one at right requires a wiring diagram for testing and the one at left has self-locking terminals.

Control thermostat
A switch that regulates a heating element; it closes or opens the circuit in cycles in response to temperature, according to the setting of its control knob. This one also has a tipover weight, which opens the circuit if the appliance falls over.

Overload protector
A thermostat switch that cuts off current in a circuit when it senses a specific amount of heat.

Humidistat
A switch that closes or opens the circuit in cycles in response to humidity, according to the setting of its control knob.

Momentary switch
Returns to its normal position once its activating button or lever is released. May be normally open (NO) or normally closed (NC); often activated by the closing of a door or the lowering of a float.

Single pole, single throw switch
Opens and closes a circuit. May have a toggle, rocker or button; typically an on/off control.

Testing switches. Switches route, vary or interrupt the electrical current through a circuit. They may be activated by hand, by the action of a moving part inside the appliance or by a change in temperature or humidity. Some typical small-appliance switches are shown at left.

Switches that control only one circuit, such as an on/off switch, can be tested with a continuity tester or a multitester set at RX1 *(page 113)*. Disconnect at least one of its wires to take the switch out of circuit, and test for continuity between its terminals. Change the position of the switch's button or toggle and repeat the test. A properly operating switch will show continuity in only one position.

Multiple-position switches control more than one circuit. They have several terminals and contacts. Without a wiring diagram, it is sometimes impossible to know when the terminals should show continuity and when they should not. If you suspect a faulty switch, first repair any loose or broken wire connections *(page 118)* and try cleaning the switch as described below. If the switch still doesn't work, replace it.

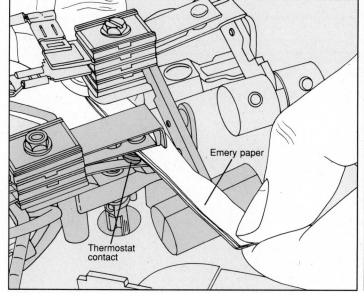

Emery paper

Thermostat contact

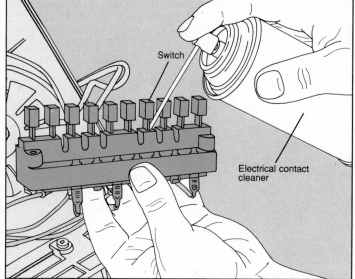

Switch

Electrical contact cleaner

Cleaning switches. If you suspect a dirty or faulty switch, unplug the appliance and consult its chapter to disassemble only what is necessary to reach the switch. A visual inspection of the switch will often reveal the problem; check the terminals for burn marks, and the casing for built-up food or dirt. If the switch contacts are visible, check whether they close cleanly and firmly. To remove dirt or smooth pitted contacts, draw a piece of fine emery paper, folded in half, between the contact leaves *(above, left)* and follow up with a piece of plain paper to pick up any grit. If the contacts are concealed within the switch casing, spray electrical contact cleaner into each aperture. Sometimes, as shown here, it may be easier to remove the switch first. Operate a control switch as you spray to work in the cleaner *(above, right)*. The control buttons on blenders and other kitchen appliances often get stiff from caked-on food; use a damp foam swab to clean around them. If the buttons continue to balk, replace the switch with an exact replacement from an authorized service center.

TESTING AND REPLACING WIRES AND CABLES

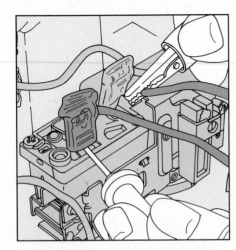

Testing wiring. If you suspect a faulty power cord, test it *(page 116)*. Check for burned or loose connections. To test internal wiring, set a multitester to RX1 *(page 113)*. Disconnect or de-solder *(page 120)* one end of the wire from its terminal and touch a tester probe to each wire end *(above)*. Flex the wire. The multitester should register continuity. If not, replace the faulty wire *(right)*.

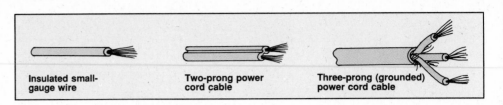

| Insulated small-gauge wire | Two-prong power cord cable | Three-prong (grounded) power cord cable |

Choosing the right wire or power cord. Most small appliance wiring is made of many copper or aluminum strands wrapped together and sheathed in plastic, rubber or heat-resistant insulation. Occasionally, solid uninsulated wires called bus wires are used. A three-prong power cord has an insulated ground wire *(above)*. When replacing a wire or power cord, choose one of the correct gauge, or thickness, indicated by a number usually printed on the insulation *(below)*. The smaller the number, the thicker the wire and the more current it can carry. Match the amperage rating of the appliance with that of the wire. To determine the appliance's amperage, divide its wattage by 120 volts. The number of watts used by the appliance will be marked next to the model number *(page 112)* on the unit's base or side. Electrical parts suppliers stock a wide assortment of wires and cables. When in doubt about the gauge of wire required, snip off a length of old wire and take it with you. Never install wire or cable of a smaller gauge, and make sure that wire for a heating appliance has heat-resistant insulation. Install wire connectors on the new wire as shown below.

WIRE GAUGE AND AMPERAGE

Wire Gauge	No. 12	No. 14	No. 16	No. 18	No. 20
Amperes	20	15	10	7	5

REPAIRING WIRE CONNECTIONS

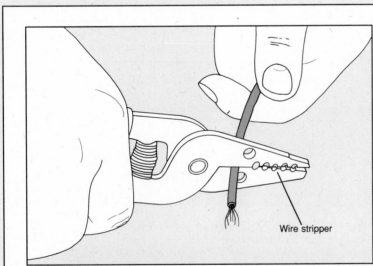

Wire stripper

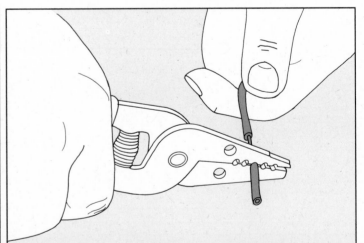

Stripping wire insulation. A wire stripper removes insulation cleanly without damaging the wire inside. It has graduated notches that accommodate the standard wire gauges, and a wire cutter near the joint. When a wire end is frayed or a connector damaged, use the wire stripper's cutting edge to snip off the damaged end *(above, left)*. If you plan to install a terminal connector, measure the length of the connector's shank and insert that length of wire into a matching notch on the stripper. If the wire is to be soldered, insert 1/4 inch of wire into its matching notch on the stripper. To strip off the insulation, close the tool and twist it back and forth gently until the insulation is severed *(above, right)*, then pull it off the wire. Before installing a wire-to-terminal connector, tin the wire lead *(page 119)*.

REPAIRING WIRE CONNECTIONS (continued)

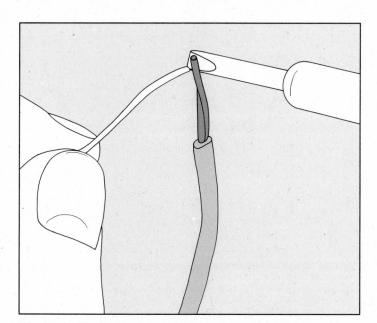

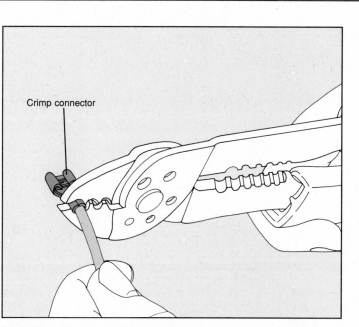

Tinning a wire lead. To prepare a wire end for tinning, strip back the insulation *(page 118)*. Prepare a soldering iron *(page 120)*. Support the wire in a soldering stand and twist the strands together clockwise between your thumb and forefinger. Hold the soldering iron tip against the wire; after a few seconds, touch the solder to the wire *(above)*. Apply just enough solder to coat the wire strands evenly. Snip off any protruding untinned wire strands using wire cutters. After tinning the lead, solder the wire *(page 121)* or install a wire connector.

Installing wire-to-terminal connectors. Most small appliance wires are connected to terminals with solderless crimp connectors. Buy the right connector for the terminal from an electrical parts supplier; if the old connector was insulated, the new one should be, too. Before installing a connector, strip off the wire insulation *(page 118)* and tin the wire lead *(left)*. Slip the connector over the tinned wire lead so that the edge of the wire's insulation butts the shank of the connector. Crimp the connector using the correct notch on a multipurpose tool *(above)*.

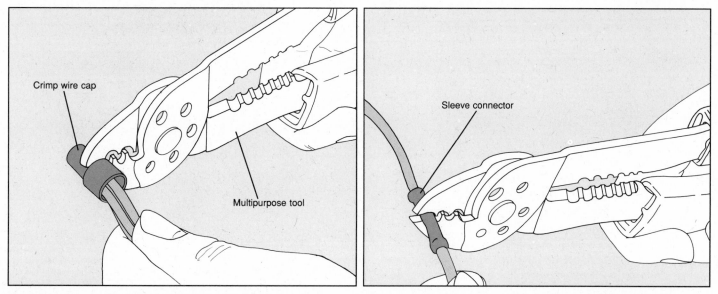

Installing wire-to-wire connectors. Sometimes two or more wires in a small appliance are joined in a crimp wire cap. When a wire connection works loose, or after testing or replacing one of the wires, you must install a new wire cap. Strip the insulation off the wire ends *(page 118)* and twist them together. Slide the wire cap over the wire ends until no wire is exposed. Fit the jaws of the multipurpose tool around the wire

cap, just above the collar, and squeeze the handles together *(above, left)*. Test the connection with a slight tug.

When joining two wires with a sleeve connector, strip 1/2 inch of insulation from each wire and slip them into opposite ends of the connector. To secure the connection, use the crimping jaws of a multipurpose tool to squeeze each end of the connector *(above, right)*. Test the connection with a slight tug.

PREPARING A SOLDERING IRON

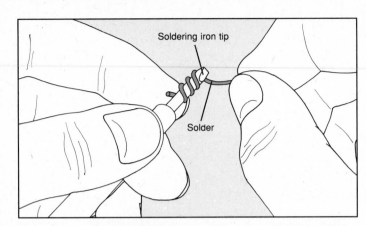

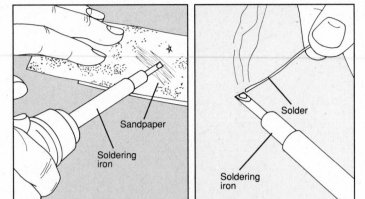

Tinning a new soldering iron. To ensure precise and tarnish-free soldering, the tip of a new soldering iron must be tinned, that is, coated with solder. Set up a soldering iron stand and place a dampened sponge on its base. Wrap a short spiral of rosin-core solder around the soldering iron tip *(above)*; use just enough to cover the entire tip evenly. Plug in the soldering iron and let it melt the solder; turn the iron's handle to help spread the solder evenly. Lightly wipe the tip across the sponge to distribute the solder over the entire tip and to remove excess. Keep the soldering iron in its stand when not in use and unplug it when you take a break.

Preparing a used soldering iron tip. If the soldering iron tip is pitted or blackened from use, rub it back and forth on a piece of medium-grade sandpaper *(above, left)* until the copper shows through. Set up a soldering iron stand and place a dampened sponge on its base. Plug in the iron and allow it to heat. Hold rosin-core solder against the sanded area *(above, right)* until it melts. Turn the iron's handle to help spread the solder evenly, then lightly wipe the tip across the damp sponge. Keep the soldering iron in its stand until you are ready to use it and unplug it when you take a break.

DESOLDERING

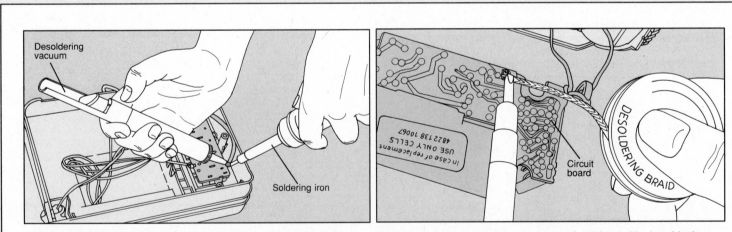

Removing old solder. Prepare the soldering iron *(steps above)*. Lightly press the tip of the iron directly on the old solder for a few seconds, until it begins to melt; if the iron is held any longer, the heat generated may damage nearby components. To remove large deposits of solder, use a desoldering vacuum. Push down the plunger, position its nozzle directly over the molten solder and use your thumb to press the trigger *(above, left)*, sucking up the melted solder. Repeat the procedure until all the solder has been removed from the terminal and the wire connection is loosened. Use long-nose pliers to remove a wire wrapped around a terminal. Remove solder from a circuit board, or clean up stubborn traces of solder, with desoldering braid. Place the braid over the old solder, gently touch the tip of the iron to the braid for a few seconds *(above, right)* and pull up the braid. Repeat the procedure until all the old solder is absorbed, cutting away used braid when necessary. Clean up accidental solder drops by desoldering them the same way. After desoldering, wipe the tip of the soldering iron on the sponge, and unplug it from the wall outlet. To clean off flux residue left by old solder, rub it gently using a foam swab moistened with denatured alcohol or with flux-cleaning solvent, available from an electronics parts supplier.

SOLDERING

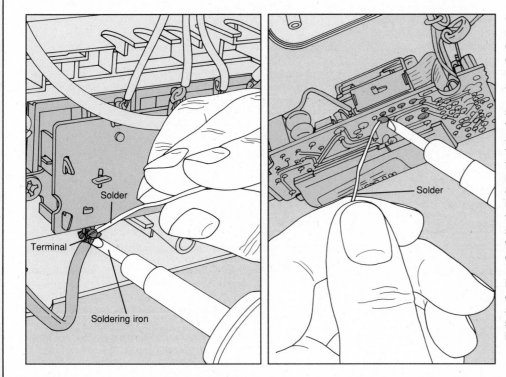

Connecting parts with solder. If you have never used a soldering iron, practice soldering lengths of scrap wire or discarded components. Prepare a soldering iron and remove old solder *(page 120)*. If soldering a wire, strip back the insulation *(page 118)* and tin the wire lead *(page 119)*. To solder a wire to a terminal, support the wire against the terminal or wrap it around the terminal. Touch the iron's tip to the wire for a few seconds and touch the solder to the heated wire *(far left)*; do not let the iron and solder touch each other directly. Keep the iron in position until enough solder has melted to secure the joint; the soldered joint should look smooth and rounded. While soldering, wipe the iron's tip on the sponge frequently. If soldering on a circuit board *(near left)*, keep in mind that less solder is needed than for a typical electrical connection, and take special care to avoid touching nearby components. After soldering, wipe the tip of the soldering iron on the sponge, and unplug it.

SILVER SOLDERING

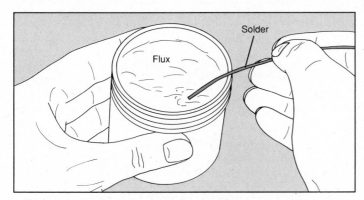

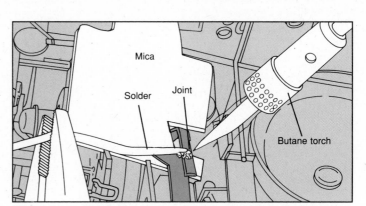

1 **Preparing to silver solder.** To replace a spot-welded electrical connection, use silver solder and a miniature butane torch. If you have never soldered before, practice first with rosin-core solder and a soldering iron *(step above)*, and then with silver solder and a butane torch. Purchase silver solder and flux paste from a welding supply store, and a miniature butane torch from an electronics parts supplier. Prepare to work in a well-ventilated area. Position the joint as securely as possible and mask adjacent components with a piece of ceramic tile or mica. Wearing safety goggles, dip the solder into the flux *(above)*, taking care not to let the flux touch your skin. Spread flux on the joint with a foam swab.

2 **Silver soldering.** Set the solder aside while you light the butane torch. Adjust it according to the manufacturer's instructions, until the flame is cone-like with a bright blue center. Pick up the solder with long-nose pliers and position it near the joint, while you heat the joint with the blue part of the flame *(above)*. As soon as a small puddle of solder spreads over the joint, withdraw the solder and the flame. Turn off the torch. To check that the connection is solid, squeeze it with the jaws of the pliers. Wipe away flux residue using a foam swab moistened with hot water. After storing the flux, solder and torch, wash your hands thoroughly.

SERVICING RECHARGEABLE APPLIANCES

Troubleshooting a rechargeable appliance. Many small appliances, such as shavers, hand mixers, electric knives and hand vacuums, are designed so that they can be used without a power cord. To do this, they are supplied with two or more special nickel-cadmium, or nicad, batteries, which must regularly be charged to full capacity for efficient operation. A charger circuit transforms household AC current to DC current that is stored by the nicad batteries and used by the small DC motor in the appliance. In an all-in-one rechargeable appliance, the charger circuit and the nicads are contained in the body of the appliance. A stand-type rechargeable appliance has the charger circuit in the stand and the nicads in the body of the appliance. Read the instructions supplied with the appliance and make sure that you charge the nicad batteries to full capacity; undercharging will shorten their life. When a rechargeable appliance doesn't work, first consult the Troubleshooting Guide in the chapter for that appliance. If you cannot locate the problem, suspect the nicad batteries. The major symptom of worn nicads is that the appliance operates for shorter and shorter periods of time after charging.

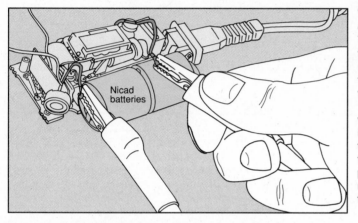

Testing the charger circuit in an all-in-one appliance. Turn off and unplug the appliance and remove its detachable cord. Take apart the appliance *(page 115)* and lift out the nicad batteries. To keep track of which wire goes to which nicad terminal, place a mark on the nicads near each terminal indicating the color of its wire. Then use a soldering iron to desolder the wires *(page 120)*. Set a multitester to test DC voltage at the 10-volt setting *(page 113)*, and clip a probe to each desoldered wire *(left)*. Reattach the power cord to the appliance, then plug it into a wall outlet without touching the appliance. The multitester should indicate 1.5 volts or more. If the needle dips to the left, unplug the appliance, reverse the tester probes and plug it in again. If the charger circuit tests OK, suspect worn nicad batteries and replace them *(below)*. If the charger circuit tests faulty, take the appliance for professional service.

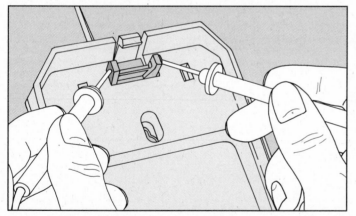

Testing the charger circuit on a stand-type appliance. Remove the appliance from the stand and locate the charger circuit terminals, usually inside the stand or on its front. Set a multitester to test DC voltage at the 10-volt setting *(page 113)*. Plug the stand into an outlet. Without touching the stand, touch an insulated tester probe to each charger terminal *(left)*. The multitester should register at least 1.5 volts. If the needle dips to the left, reverse the probes. If the charger stand tests OK, suspect worn nicad batteries in the appliance and replace them *(below)*. If it tests faulty, take the stand for professional service.

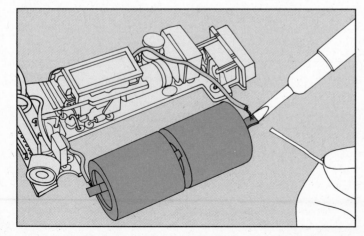

Replacing nicad batteries. Disassemble a stand-type appliance and remove its nicad batteries as in the first step above. Take the nicad batteries to an authorized service center and buy exact replacements. Solder *(page 120)* the wires to the terminals of the new nicads, using the marks you made on the old nicads as a guide *(left)*. Reinstall the nicad batteries and reassemble the appliance, reversing the sequence of steps you took to disassemble it. Plug in the appliance and charge the nicad batteries to full capacity. If the appliance won't charge, you have installed the nicads backwards. Disassemble the appliance, reverse the nicads and reassemble it again.

TROUBLESHOOTING GUIDE SMALL APPLIANCE MOTORS

An electric motor changes electrical energy into mechanical energy. When an electrical current runs through a motor, two opposing electromagnetic fields are set up, one around the movable rotor windings and one around the stationary field windings. The interaction of the two fields spins the rotor.

The shaded pole motor and the universal motor pictured below are those most frequently found in small appliances. The simpler shaded pole motor is identified by its two pairs of heavy copper wires, called shading coils. It operates on alternating current only and is used in light-duty appliances such as can openers and fans. Heavy-duty appliances such as vacuum cleaners and stand mixers require the power of a universal

motor. So named because it can operate on either alternating or direct current, the universal motor is identified by its exposed copper coils and by the two brushes (actually small bars of carbon) that press against a cylinder of brass bars called the commutator. The bearings, brushes and commutator may be serviced, but most small-appliance motors that test faulty must be replaced. Review the Troubleshooting Guide below, and refer to pages 124 and 125 for motor testing and repair procedures. Before starting any repair, discharge the capacitor *(page 114)*, then inspect the wires and install new connectors if necessary *(page 118)*. Purchase replaceable motor parts from the appliance's authorized parts supplier or from the manufacturer.

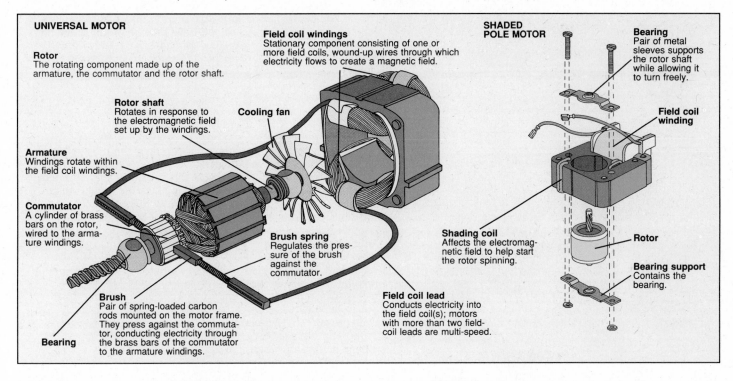

UNIVERSAL MOTOR

Rotor
The rotating component made up of the armature, the commutator and the rotor shaft.

Rotor shaft
Rotates in response to the electromagnetic field set up by the windings.

Armature
Windings rotate within the field coil windings.

Commutator
A cylinder of brass bars on the rotor, wired to the armature windings.

Brush
Pair of spring-loaded carbon rods mounted on the motor frame. They press against the commutator, conducting electricity through the brass bars of the commutator to the armature windings.

Bearing

Field coil windings
Stationary component consisting of one or more field coils, wound-up wires through which electricity flows to create a magnetic field.

Cooling fan

Brush spring
Regulates the pressure of the brush against the commutator.

Field coil lead
Conducts electricity into the field coil(s); motors with more than two field-coil leads are multi-speed.

SHADED POLE MOTOR

Bearing
Pair of metal sleeves supports the rotor shaft while allowing it to turn freely.

Field coil winding

Shading coil
Affects the electromagnetic field to help start the rotor spinning.

Rotor

Bearing support
Contains the bearing.

SYMPTOM	POSSIBLE CAUSE	PROCEDURE
Motor doesn't run at all	Brushes not contacting commutator	Replace brushes *(universal motor, p. 124)* ▤○
	Commutator grounded to rotor shaft	Test for a ground fault *(universal motor, p. 125)* ▤○▲
	Open circuit in motor windings	Test motor *(universal or shaded pole motor, p. 125)* ▤○▲
Motor sparks, sputters, runs sluggishly or vibrates severely	Brushes worn, poor contact with commutator	Replace brushes *(universal motor, p. 124)* ▤○
	Commutator or rotor dirty	Clean commutator *(universal motor, p. 124)* ▤○ or rotor *(shaded pole motor, p. 125)* ▤○
	Short or open circuit in armature windings	Test motor *(universal motor, p. 125)* ▤○▲
	Mica protruding through commutator	Scrape mica off commutator *(universal motor, p. 124)* ▤○
Motor sparks, overheats, has burning odor	Short circuit in motor windings	Test motor *(universal motor, p. 125)* ▤○▲
Rotor shaft turns with difficulty	Bearings binding or rotor dirty	Clean and lubricate bearings *(universal motor, p. 124; shaded pole motor, p. 125)* ▤○
Rotor shaft too loose or doesn't turn at all	Bearings seized or worn	Take appliance for professional service

DEGREE OF DIFFICULTY: □ Easy ▤ Moderate ■ Complex
ESTIMATED TIME: ○ Less than 1 hour ◑ 1 to 3 hours ● Over 3 hours   ▲ Special tool required

SERVICING A UNIVERSAL MOTOR

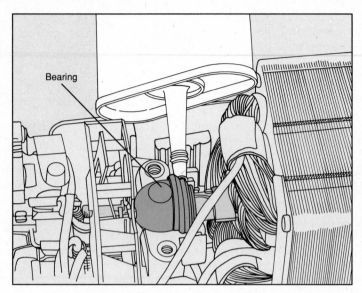

Cleaning and lubricating motor bearings. Unplug the appliance. Refer to the appliance's chapter and to the tips on disassembly *(page 115)* to access the motor. Turn the rotor shaft manually; if it doesn't budge at all, or if it has excess play in the bearing, take the appliance for service. If the shaft moves stiffly, the bearings may be binding and need lubrication. Usually there are two bronze or soft-metal sleeve bearings on the rotor shaft of a small appliance's universal motor, one at each end of the motor. They may be hidden; one bearing on the mixer motor shown here is normally concealed by a bracket, the other, by the gear box. Remove any bracket or gear box cover that hides the bearings. Clean off old lubricant with a foam swab dampened with denatured alcohol; apply the alcohol to the shaft and bearing as you rotate the shaft. Then carefully apply two drops of non-detergent 20-weight machine oil *(left)*. Don't get oil on the commutator or brushes. Reassemble the appliance, reversing the steps taken to access the bearings, and cold check for leaking voltage *(page 114)*.

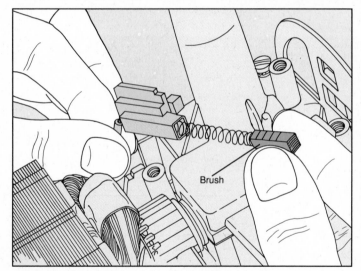

Cleaning and replacing motor brushes. Unplug the appliance. Refer to the appliance's chapter and to the tips on disassembly *(page 115)* to access the motor. Check both motor brushes. To remove a brush for inspection, unclip or unscrew the brush housing, if necessary. Be prepared to catch the brush, which may fly out on the end of its spring *(left)*. Some brushes can be unclipped from the rear and pulled out by the spring, leaving the housing in place. Inspect the brush, the spring and the brush housing. If the brush or housing is dirty, spray it with electrical contact cleaner. If the spring is broken or damaged, replace it. If either brush is shorter than it is wide, replace both brushes. Buy exact replacement springs and brushes from a small appliance service center that specializes in your appliance. Install brushes with the curved end aligned correctly with the curve of the commutator. Reassemble the appliance, reversing the steps taken to access the brushes, and cold check for leaking voltage *(page 114)*.

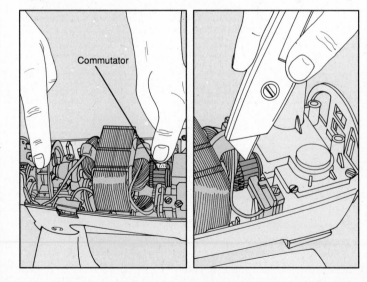

Cleaning the commutator. Unplug the appliance. Refer to the appliance's chapter and to the tips on disassembly *(page 115)* to access the motor. Rotate the commutator by turning the rotor shaft, and look for evidence of burning, pitting or dirt buildup. To clean the commutator, use your fingers to press a small piece of emery paper lightly against the bars as you rotate the shaft *(far left)*. Use an old toothbrush to clean oxidation or dirt from between the bars. If the mica insulation protrudes above the brass bars, use a utility knife to scrape it from between the bars, until it is slightly below the level of the bars *(near left)*. Reassemble the appliance, reversing the steps taken to access the brushes, and cold check for leaking voltage *(page 114)*.

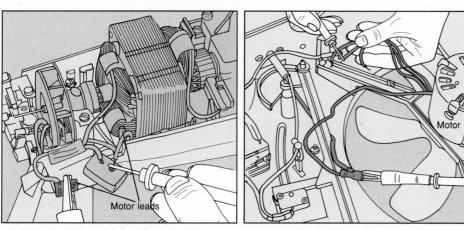

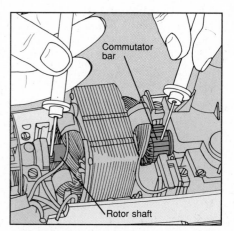

Testing the motor. Unplug the appliance. Refer to the appliance's chapter and to the tips on disassembly *(page 115)* to access the motor. Set a multitester to RX1 *(page 113)*. To test a single-speed or variable-speed motor, disconnect one motor lead and touch a tester probe to each lead *(above, left)*. Rotate the rotor shaft as you test. The multitester should show low resistance. To test a multi-speed motor, locate the speed control switch and label and disconnect the motor leads from it. Locate the motor lead not connected to the switch and disconnect it also. Clip a probe to this lead and touch the other probe to each of the other leads, in turn *(above, right)*, as you or a helper rotate the rotor shaft. The multitester should show a different resistance reading for each lead. If the motor tests faulty, take the appliance for professional service.

Testing for a ground fault. Unplug the appliance and access the motor. To test whether the commutator is grounded to the rotor shaft, set a multitester to RX100 *(page 113)*. Touch one probe to the rotor shaft and touch the other to each of the commutator's bars, in turn *(above)*. The multitester should show infinite resistance for each bar. If a ground fault is detected, take the appliance for professional service.

SERVICING A SHADED POLE MOTOR

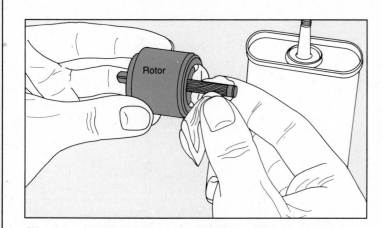

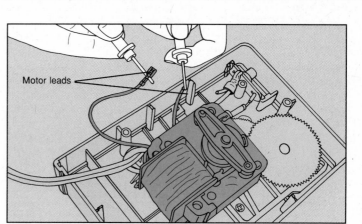

Cleaning and lubricating a shaded pole motor. Unplug the appliance. Refer to its chapter and to the tips on disassembly *(page 115)* to access the motor. Disassemble the motor by locating and removing the bearing supports. Pull out the rotor. Use a foam swab dipped in denatured alcohol to clean the rotor shaft and then lightly oil the shaft with a small amount of non-detergent 20-weight machine oil *(above)*. Lightly lubricate the bearings and wipe away excess oil. Slide the rotor back into its opening. Resecure the bearing supports and reassemble the appliance, reversing the steps taken for disassembly.

Testing a shaded pole motor. Unplug the appliance. Refer to its chapter and to the tips on disassembly *(page 115)* to access the motor. To test the field winding of a shaded pole motor, set a multitester to test resistance *(page 113)*. Disconnect one of the motor's leads. Touch a tester probe to each lead *(above)*. The multitester should indicate low resistance. If the motor tests faulty, take the appliance for professional service.

INDEX

Page references in *italics* indicate an illustration of the subject mentioned.
Page references in **bold** indicate a Troubleshooting Guide for the subject mentioned.

ACKNOWLEDGMENTS

The editors wish to thank the following:
John Aiello, Authorized Appliance of Westchester, Tuckahoe, N.Y.; Stewart Atkin, Montreal, Que.; Louise Bedard, Singer Sewing Machine Company of Canada Limited, St. Jean, Que.; Patrick Benfante, Benfante Appliance Service, Rochester, N.Y.; Gregory Brent, Montreal, Que.; Sharon Brucken, Sunbeam Appliance Service Company, Downers Grove, Ill.; Archie Burns, Don Reedy Appliance Service, Silver Spring, Md.; Danny Carlisle, Sutton, Que.; Dan Carozza, Lake Electronic Service Inc., Albany, N.Y.; Jean Charette, Vacuum Rebuilders of Montreal, Montreal, Que.; Vaclav Chlumsky, Stancan Reg'd., Montreal, Que.; Robert Claus, Bemis Manufacturing Company, Sheboygan Falls, Wis.; Darrell Darnell, Greenville, S.C.; Christian Demers, Biotech Electronics Limited, Lachine, Que.; Yves Deschesne, Montreal, Que.; Jerry Duplessis, Mr Fix-It Service Center, Montreal, Que.; Lorne K. Fortin, Black & Decker Canada Inc., Brockville, Ont.; Stephen Gilson, St. Lambert, Que.; Julio Gomez, Spunt's Radio Service, Montreal, Que.; Lawrence J.B. Hampton, Toronto, Ont.; Paul Holroyd, Delonghi Canada, Mississauga, Ont.; Barry Inglis, Singer Sewing Machine Company of Canada Limited, St. Jean, Que.; Michael Jaggi, Ottawa, Ont.; Philippe Lacombe, Sunbeam Appliance Service, Montreal, Que.; Pierre Marchetti, BM Electronique, St. Leonard, Que.; Wayne McGill, Mr Fix-It Service Center, Montreal, Que.; Cynthia Monsky, Singer Sewing Company, Edison, N.J.; Tyrus S. Peele, Ace Hyattsville Sew-Vac, Hyattsville, Md.; Don Ringstrom, Singer Sewing Company, Edison, N.J.; Albert Scheffer, Toshiba of Canada, Dorval, Que.; Harold Siegwarth, Camco Inc., Montreal, Que.; William Staples, Emerson Builder Products, St. Louis, Mo.; Mario Viel, Larocque Electrique, Montreal, Que.; Paul Widstrand, Oster, Milwaukee, Wis.; K.A. Woods, Black & Decker Canada Inc., Brockville, Ont.

The following persons also assisted in the preparation of this book:
Arlene Case, Cathleen Farrell, Patrick J. Gordon, Julie Leger, Francine Lemieux, Odette Sevigny, Natalie Watanabe, Billy Wisse, Katherine Zmetana.

Typeset on Texet Live Image Publishing System.